TRUE STORIES FROM THE BRAVEST OF THE BRAVE

AMERICAN HEROES

JAMES PATTERSON
AND MATT EVERSMANN

with **Tim Malloy**
and **Chris Mooney**

CENTURY

1 3 5 7 9 10 8 6 4 2

Century
One Embassy Gardens
8 Viaduct Gardens
London SW11 7BW

Century is part of the Penguin Random House group of companies whose addresses can be found at global.penguinrandomhouse.com.

Copyright © James Patterson 2024

James Patterson has asserted his right to be identified as the author of this Work in accordance with the Copyright, Designs and Patents Act 1988.

First published in the UK by Century in 2024

www.penguin.co.uk

A CIP catalogue record for this book is available from the British Library.

ISBN 9781529136654 (hardback)
ISBN 9781529136661 (trade paperback)

Printed and bound in Great Britain by Clays Ltd, Elcograf S.p.A.

The authorised representative in the EEA is Penguin Random House Ireland, Morrison Chambers, 32 Nassau Street, Dublin D02 YH68

www.greenpenguin.co.uk

Penguin Random House is committed to a sustainable future for our business, our readers and our planet. This book is made from Forest Stewardship Council® certified paper.

AMERICAN HEROES

In memory of Colonel Ralph Puckett, Jr.

December 8, 1926-April 8, 2024

I've conducted hundreds of interviews for our books but only one in person: Colonel Ralph Puckett, Jr., an icon in the Army and especially the Ranger world. As luck would have it, I was in Highlands, North Carolina, when he and his beloved wife, Jeannie, invited me to visit their lovely home in Columbus, Georgia. I couldn't say no. No Ranger would even dream of saying no.

It was a cool October morning and I was nervous. I was actually going to sit in Colonel Ralph Puckett's living room and ask him questions about his brilliant career and, specifically, about the battle in Korea, for which he received the Congressional Medal of Honor.

Back in the 1990s, when I first met the then-Honorary Colonel of the Ranger Regiment, it was a pretty big deal to listen to him speak—not in a lecture hall, but in the field, on the rifle range, in the middle of the night, after a parachute jump or a long road march. Colonel Puckett was right there, witnessing the training and giving guidance on what he saw.

More importantly, he talked *with* less experienced soldiers. His words weren't saccharine chatter spoken out of kindness. They were instructive and informative and made every one of us—the young Ranger privates and sergeants and lieutenants—feel like his peer. He took a personal interest in everyone—soldiers and Rangers—the warriors of America. Always.

When we spoke in his apartment that October day, he mentioned that he was looking forward to "getting back to post" to visit with the next generation of Army leaders. At age ninety, he was slowing down a little, but when we finished the interview, he stood to shake my hand. The Pucketts invited me to stay for lunch—barbecue from a local place—and more lively conversation. We debated the best hot sauce and enjoyed cookies, shaped like Ranger tabs, of course, for dessert. Throughout the visit we spoke as if we had known each other forever.

That's the man Ralph Puckett, Jr. was: a decorated soldier, a selfless servant, a gentleman. You'll read about that battle during the Korean War, and his story, like every story in this book, will leave you in awe. When I think about the order he gave to his men to leave him on the hill, I know without a doubt that he was a man who would unhesitatingly give such an order. Maybe for that exact reason, his Rangers disobeyed the order and brought their commander and Ranger Buddy to safety. He was a treasure then and a treasure now. We all were blessed to have known him.

Rest in Peace, Sir. Rangers Lead the Way.

—M.E.

Contents

Hershel "Woody" Williams	1
Patrick Henry Brady	14
Paris D. Davis	28
Thomas William Bennett	42
Cindy Pritchett	53
Ralph Puckett, Jr.	68
Duane Edgar Dewey	82
Alwyn Cashe	94
Harvey Curtiss "Barney" Barnum, Jr.	104
Matthew O. Williams	116
Donna Barbisch	124
David G. Bellavia	137
Chuck Pfeifer	148
Paul Zurkowski	157
Arthur Rice	167
Michelle Saunders	174
Mark Mitchell	186

CONTENTS

Brian M. Kitching	206
Conrad Begaye	220
McKenna "Frank" Miller	226
Neil Prakash	242
Tim Sheehy	254
Jay Zeamer, Jr.	266
Earl D. Plumlee	270
Alan Mack	289
Gary Wetzel	304
Charles Ritter	316
James Livingston	334
Travis Mills	346
Ernest E. West	361

Hershel "Woody" Williams

Corporal, US Marine Corps Reserve
Conflict/Era: World War II
Action Date: February 23, 1945
Medal of Honor

Interview with Brent Casey, grandson

My grandfather, Hershel "Woody" Williams, was a Marine who fought in World War II. A picture of Jesus hangs on the wall of his farmhouse in Ona, West Virginia. Two vials of sand from Iwo Jima sit on a kitchen shelf near a picture of him dressed in his OD-green Marine uniform shaking the hand of President Truman as he received the Medal of Honor.

As a kid, I don't know the details of what my grandfather did during the war or why he was awarded the Medal of Honor. I don't know anything about the medal or its significance, only that he doesn't wear the medal often. Maybe to a local parade or when he and my grandmother travel to some event or gala.

JAMES PATTERSON

If my grandfather wants something done, he only asks once. He works for Veterans Affairs (VA), where he's respected and admired enormously. He also has a horse farm, where at any given time he's raising and training from ten to fifteen horses for other people to show.

Woody isn't opposed to using corporal punishment to get a horse—or a kid—back in line. I'm the only grandchild who hasn't received a horse crop to my backside, but when, at sixteen, I'm having some behavioral challenges, my mom's solution is to send me to work on my grandfather's farm.

"He's going to tell you how things are supposed to be and show you how things are supposed to be," she says.

Woody teaches me a lot about selflessness and how it relates to discipline and hard work. He inspires me to serve my country.

At nineteen, I join the Army. That's where I learn about my grandfather's actions in Iwo Jima. Why he's a hero who played a part in protecting this great country.

Woody is seventeen years old and lives on a dairy farm in Quiet Dell, West Virginia, with his mother and eleven siblings. War is coming—one, he is sure, that will take away America's freedoms and privileges.

On October 2, 1923, Woody was born on the farm weighing three and a half pounds. Mom lost several children to the 1918–1919 flu pandemic and was sure she was going to lose him, too.

The country was in the grips of the Great Depression when Woody's father died of a heart attack. Eleven-year-old Woody helped support his family, working as a taxi driver and delivering telegrams to the families of American soldiers who were killed during the early days of World War II.

Marines wear snappy blue uniforms around town and Woody is impressed. If he's going to war, he's going as a Marine. At seventeen, he needs his mother's signature to enlist.

Mom refuses to sign the papers. He turns eighteen next October. If he wants to be a Marine, he's going to have to wait.

December 7, 1941: Japan attacks Pearl Harbor.

November 1942: Woody walks into the Marine Corps recruiting office. The recruiter doesn't so much as glance at his papers. The man is too busy studying Woody's stature.

"I can't take you," the recruiter says. "You're too short."

To be a Marine, Woody is told, you must be five eight or better. He's five six.

His two brothers have been drafted by the Army. He might've seriously considered following in their footsteps if it weren't for the Army's old brown wool uniform. It's the ugliest thing in town.

I want to be in those dress blues, he thinks.

If he can't be a Marine, well, he's not going. Woody returns to working on the dairy farm with Mom.

In early 1943, the recruiter shows up on his doorstep. The Marine Corps, he explains, has lowered the height requirement.

"Do you still want to go?"

"Yes," Woody replies, and signs up.

When he thought about protecting his country, he believed it would happen here, on US soil. Instead, he's sent eight thousand miles from home. On the boat, he looks at a map of the pork-chop-shaped island of Iwo Jima.

His unit is in reserve to two other Marine divisions for a campaign expected to last three to five days. There's no intelligence, only belief in the strong chance that his unit will never leave the ship. He doesn't know—no one does—that twenty-two thousand Japanese have dug miles of tunnels across the island and entrenched there.

The Marines sustain catastrophic losses on the beach. Woody and his unit are parked out at sea. That night, they receive new orders to board Higgins boats to Iwo Jima.

There's no room for them to land. All day long, they huddle down and circle, awaiting the signal. It never comes.

The next day, they head into the beach.

The ramp lowers and Woody pops his head up, catching a glimpse of chaos. The beach is strewn with bodies and limbs torn apart by bullets and explosions; blown-up tanks and Jeeps and equipment; craters and burnt vegetation.

Exiting the ramp, Woody does a double take. Dead Marines are wrapped in rain ponchos, their bodies stacked like cordwood along the black sand beach made from volcanic ash.

Woody and the six Marines with him crawl across fine sand the texture of BBs. Dig a hole and the sand refills it. Near the first airfield, sand becomes soil where they can dig foxholes.

A flag! A flag!

The Marines start yelling and shooting their rifles into the air. Then Woody sees it—an American flag mounted on Mount

Suribachi, the wind unfurling its cotton fabric appliquéd with the familiar red, white, and blue stars and stripes. He joins the celebration and fires his rifle.

The airfield is protected by a cluster of bazooka-proof concrete pillboxes reinforced with rebar and fronted with a four-inch-tall aperture where the Japanese place the barrells of their Nambu machine guns and rifles to mow down US soldiers.

An all-NCO meeting is called. Woody is a corporal, not an NCO, but Woody's captain tells him, "I want you there."

Sheltered inside a huge shell crater, the commanding officer explains the grim reality. The Marines are in an open area. Unless the pillboxes are taken out, there's no way to advance against the Japanese in their protected position. They'll have to advance on the pillboxes without any cover.

The commanding officer looks at Woody. "Do you think you can do something with the flamethrower against some of these pillboxes?"

Woody is the only flamethrower left in C Company. The other Marines who came with him, he has no idea where they are, if they're alive or dead. He'll never know, even decades later.

"I'll try," Woody says.

He trained as a demolition man stateside, knows how to use the M2 flamethrower. Its two fuel canisters filled with compressed gas and liquid diesel fuel and high-octane gasoline are mounted on a steel rack fitted as a backpack and altogether weighing close to seventy-five pounds. A full blast lasts about seventy-two seconds. The weapon's average lifespan: four minutes.

The key is to pull the trigger in bursts. Once the canisters run dry, the operator has to roll out of the backpack and put on a new one.

He'll be moving slowly behind a bright orange flame that will draw the attention of enemy snipers.

He's scared to death, but reminds himself, *Everyone has some instinct of bravery. As long as you can control the fear, you can be brave.*

He thinks of Ruby, the girl waiting for him back home. When the war is over, he's going to marry her.

Four Marines are assigned to cover Woody. He selects another Marine to be his "pole charge man." After Woody sprays liquid fire through the aperture, the pole charge man probes the pillbox with a long piece of wood capped with an explosive fatal to any enemy who's survived the flamethrower attack.

It's time to get to work.

The combat is fierce, so intense as to blur his actions over the next four hours. Some events and moments he'll never recall. Others will live vividly within him forever.

He remembers crawling on his stomach to reach the top of one pillbox. There, he sticks the flamethrower's nozzle down a vent pipe and squeezes the dual trigger. He remembers his pole charge man taking a round to the head. The man's helmet saves his life, but he's taken out of commission. Woody takes over the man's duties preparing the explosives.

He has no idea how he repeatedly gets his hands on fresh fuel canisters—people say he ran back to the lines; others say Marines delivered canisters to him. Two of the four Marines assigned to cover him are killed. He remembers how each man died and he remembers being on his own, bullets pinging off the tanks strapped across his back and, near the end, Japanese soldiers bursting out of the pillbox with bayonets fixed to the rifles, determined to take down the small American Marine who has killed so many of their own. Woody remembers, will always remember, pulling the trigger and watching the Japanese soldiers catch on fire before falling to the ground.

He feels no remorse. The Japanese have killed many Marines — including his best friend, Vernon Waters. The battle — the war — needs to be won. He keeps fighting, a one-man killing machine, and by the time he's done, he has managed to wipe out seven pillboxes.

Woody doesn't think he's done anything special. He was just doing his job.

In the Army's 82nd Airborne Division, I serve as a combat medic deployed to Saudi Arabia to support Operation Desert Shield/Desert Storm. Day after day, I try to do my job while another part of me is full of fear — fear of the unknown.

In 1991, I return home. That's not true. Only my body returns. My brain... is still somewhere overseas.

I get divorced and hit the bottle. I self-medicate for fifteen years. I end up homeless, jobless, and carless. I have absolutely nothing left except for addiction and alcoholism.

In 2005, my mother, brother, and Woody come collect me. I'm taken to the VA hospital in Huntington. I've never been to the VA. The doctor tells me that I've flagged 19 out of 19 on my PTSD questionnaire.

"What the hell is PTSD?" I ask.

"Don't worry about that," the doctor replies. "You're in the right place. We've got all the tools and resources you need to get back on track."

I get involved in the VA's rehab and PTSD residential rehabilitation program. I learn healthy coping skills. Woody helps me the entire time. He, too, was diagnosed with PTSD. He's all too familiar with my struggles.

My grandfather shares the nightmares he had after the war—fighting fires instead of shooting fire. One time he jumped out of bed and pounded the wall because he believed it was on fire.

Attending the Pea Ridge Methodist Church changed his life. The nightmares stopped, but the regrets he had over killing people remained.

The VA completely changes my life. I go back to school, get a bachelor's degree in business administration and management, then work full-time on my MBA with the goal of becoming a teacher.

My grandfather retires—from the VA and from horse farming. When Woody turns eighty, my grandmother says, "It's time to hand up the horse reins." He does, moving about a thousand yards down to the bottom of the hill, where he can still see the farm and, whenever he wants, walk back up and take a look at the horses.

Woody continues his speaking engagements and I often accompany him. My grandfather doesn't need the money he earns, so I say, "Papa, why don't we do something with the honorariums and create a nonprofit?"

Woody loves the idea.

I get together with my brother Bryan, who has a law degree. He works on the 501(c)(3) form while I go about assembling a board of directors for the foundation we create together. I'm building out the website when I receive a call from my grandfather.

"I was at Donel C. Kinnard Memorial State Veterans Cemetery and came up with an idea," he says. "I've been working with an architect. I'm going to snail mail you some copies of this monument I have in mind."

"This monument, what is it?"

"It's going to honor Gold Star Families."

That's what the military calls immediate family members of a soldier who died in the line of duty. I do a Google search for "Gold Star Families Memorial Monument." None exists.

"You're not going to believe this," I tell my grandfather. "Your idea for 'Gold Star Families Memorial Monument.' It's original."

Some might say—correctly—that the idea fell into our laps. But in our minds, it's a God thing, no question. It's the way Woody and my parents raised us.

The first monument is unveiled at Donel C. Kinnard Memorial State Veterans Cemetery on Woody's ninetieth birthday. The foundation starts to take off and I'm forced to make a decision. Do I want to write my PhD dissertation or run the nonprofit? I can only do one.

I decide to help my grandfather, pay him back for everything he's done for me. He's the reason I'm still alive.

The foundation dedicates twenty-five Gold Star Families

Memorial Monuments across the country. Then we start working on establishing a memorial in each state. When we reach that goal, we expand into other countries.

My grandfather insists on attending each dedication. Even at age ninety-eight and dealing with a leaky heart valve, he still moves at a hundred miles an hour. On the road together 230 days a year, we share airplane flights, hotel rooms, and meals.

One day at home, Woody makes a misstep. Instead of the bathroom, he steps into the stairwell. He tumbles down all eighteen steps and is knocked out. He breaks five ribs and his pelvis in three places.

He spends a month in the hospital, then two months inside a physical rehabilitation center. Tell my grandfather to do twenty-five push-ups and he does fifty. He's released in time for Christmas. By January, he's completely recovered.

We're in Dallas, Texas, for the groundbreaking of the National Medal of Honor Museum when we receive a call from the World War II Museum in New Orleans, Louisiana. They want to create a 4D hologram of Woody so that museum visitors from all over the world will be able to sit down and have a simulated conversation with him.

My grandfather can't wrap his head around the concept. And he isn't excited about going to California where, over several days, he'll be filmed using sophisticated technology. The week before we're scheduled to leave, he calls me to cancel.

"Papa, you don't understand," I say. "The museum has already paid for our expenses, has everything in place to film."

He doesn't want to go.

I text the other four grandsons. We all get on the phone with him, talk him through how this is a unique opportunity.

"Papa," I say to him, "my kids' grandkids will be able to go to the World War II Museum forty, fifty years from now. What better way to leave your legacy and your story behind?"

It takes some effort, but we manage to convince him. A friend offers the use of his private jet and Woody flies to Hollywood and films for four straight days.

The technology is incredible. Museum visitors will see a projection of Woody sitting in a chair. When they ask him questions about his life, what it was like growing up in West Virginia and serving overseas, Woody will answer right back.

Around March, he develops vertigo. He gets fluid in his legs. Woody ignores it, and the fluid builds until it gets inside his lungs. We take him to Huntington, West Virginia, and the same VA hospital where I was treated, now renamed the Hershel "Woody" Williams VA Medical Center.

The VA medical staff recommend further evaluation at the Cleveland Clinic. Doctors there believe Woody's heart has a second leaky valve. They can't push the fluid out, so they put him on Lasix, a strong diuretic used to treat excessive swelling. His kidneys can't handle the medication.

The doctors suggest dialysis.

"Nope," Woody says. "Nope, nope, nope. We're not doing any of that. I want to go home."

"What do you mean when you say home?" the doctor asks. "Do you want to go to a rehab center, or do you want to go to your house?"

JAMES PATTERSON

"I want to go to the Woody Williams VA Medical Center."

I need to get him home right now. It takes a lot of phone calls well into the early morning hours, but I finally arrange a StatFlight, and the company generously donates their services free of charge. I manage to get Woody on the helicopter flight home the following morning, Saturday.

By Sunday, Woody's feeling pretty good. He wants to see my mom — it's her birthday — and my youngest brother, who he hasn't seen in a while. My grandfather is in good spirits, sitting up in a hospital chair, but we all know he's sick.

Senator Joe Manchin stops by Woody's room. Woody wants the senator's help. For some time, it's been on my grandfather's mind to build a protective shelter for families at the Donel C. Kinnard Memorial State Veterans Cemetery.

"I'm going to make that happen," the senator says.

My wife, Mary, and I are with Woody in the early morning hours of June 29, 2022. I'm playing him church hymns on my iPhone. "In the Garden," Woody's favorite song, comes on, and that's when my grandfather, the last living Medal of Honor recipient who served in World War II, takes his final breath, at the VA hospital named after him.

It's another God thing.

For two days, Woody, wearing his dress blues, lies in state in the state capitol building in Charleston, West Virginia. A Gold Star Families Memorial Monument sits on the capitol lawn.

Woody's final tour continues. His casket is loaded onto a C-130 for transport to Washington, DC, to lie in honor at United States Capitol Building — the thirty-eighth American to do so.

AMERICAN HEROES

My nephew, who two years ago joined the Marine Corps because of Woody, accompanies him.

A ceremony is held in the rotunda. It's a beautiful day. Just God-given. And special. Later, Woody will be moved to the World War II Memorial on the National Mall.

My grandfather knew of these plans before his death, and only agreed once he was satisfied that the day would honor not only him but *all* World War II veterans. To his very last breath, my grandfather believed he was the caretaker of the Medal of Honor.

"I didn't earn it," he often said. "I wear it for those Marines who lost their lives protecting mine."

Patrick Henry Brady

Major, US Army
Conflict/Era: Vietnam War
Action Date: January 6, 1968
Medal of Honor

Growing up in Seattle, Washington, I attend an all-boys Catholic high school run by the Christian Brothers of Ireland. My senior year I'm offered several football scholarships. However, at the time I am stalking this foxy chick who attends an all-girls Catholic school, and she is going to Seattle University. Seattle University does not have a football team.

I have a tough decision to make. I'm broke. Free tuition or take out a loan and follow the foxy chick.

I make the best decision of my life. I follow the foxy chick.

When I arrive on campus, I discover that ROTC is mandatory! I can't believe it. I hate the military. To me that is communism. They can't make me wear that silly uniform, march, and salute.

But they can and they do. Needless to say, I don't do well.

When the time comes for Summer Camp, I decide to go to Alaska and work on the railroad. I need the money so I can marry the foxy chick. I hear that I set the record for gigs at Summer Camp before they discover I'm not there. I am then kicked out of ROTC.

I am happy to spend the rest of my life working on the railroad in Alaska, but my new wife insists I finish my education. So we go back to Seattle University. While enrolling I was recruited by Major Snyder (a great inspiration in my life), the ROTC representative. He asks if I want to finish ROTC. He knows my record.

In those days military service is mandatory. I'll be drafted or I can get a commission and go in as an officer. I have one child and another on the way. I'll make more money as an officer. My wife wants me to finish ROTC and so I do. I still don't like the military but do my best and graduate as the president of the Scabbard and Blade Society (the number one cadet), with a listing in *Who's Who among Students in American Colleges and Universities* and a Distinguished Military Graduate Award. My first assignment is Berlin, Germany, as a medical platoon leader.

It's 1959, and the East German Republic is part of the Communist bloc. Its capital, Berlin, is a divided city. Soviet troops control the eastern sector. West Berlin is controlled by World War II allies, the United States, the United Kingdom, and France.

I'm an officer—a second lieutenant—and I get to bring my family with me. It's 1959. I'm twenty-one and my wife is pregnant. We're given use of a beautiful three-bedroom apartment that's probably worth millions today. It's fully furnished, has

Rosenthal China, and a refrigerator stocked with food. There's a room for the babysitter, and I can afford a maid.

I serve with Norm Schwarzkopf and some other great officers, and while I think, damn, it wouldn't be bad to grow up and be like them, I'm still not committed to making the military my career.

One day—August 13, 1961—I wake up and the city is divided by barbed wire, which becomes the infamous Berlin Wall. They can't stop Allied forces from traveling, but they stop the Germans. Our babysitter's fiancé is in the East. She never sees him again. The maid's parents live in the East. When they die, she's not allowed to attend their funerals. Same wonderful and industrious people on both sides but living in two completely different systems, different ways of life: socialist/communist vs. American.

The differences are amazingly stark. On the east side, the people look downtrodden. The buildings are a mess. On the west side there are big, magnificent shopping centers and fancy cars. East Germans who try crossing the wall are getting shot, and my guys are tasked cleaning up the mess. It's the first time I've come face-to-face with communism.

When I leave Berlin, I go to flight school. On my twelfth birthday, I took a fifteen-minute ride in a L-19 stick plane at Seattle's Boeing Field and got hooked. I get my wings in 1963.

The United States is involved in the Vietnam War. I volunteer for Vietnam where once again I'll see communism, this time up close and personal.

* * *

Vietnam is a beautiful country, and it's hotter than hell. I'm in good physical shape, but that first month I lose seventeen pounds.

Then I meet my commander, Charles Kelly, an Army helicopter pilot. I've heard a lot about him. Kelly, an irascible Irishman, was court-martialed an amazing four times. At age fifteen, he lied about his age so he could serve in World War II (and almost died in battle). He is the only soldier to ever wear all four major Army badges (the Combat Medic Badge, the Combat Infantry Badge, the Parachute Badge, and Flight Wings). He came to Vietnam to command the 57th Helicopter Ambulance Detachment, call sign Dustoff.

Kelly is not in a good mood. He has just returned from a combat operation near the South China Sea, where a helicopter went down. Kelly was only able to rescue part of the crew. The rest drowned while he hovered over them. He's also waging a battle with his boss, Brigadier General "Cider" Joe Stillwell. Stillwell wants to put portable red crosses on Kelly's helicopters and use them part-time for nonmedical missions.

Kelly is outraged. "Unless we prove the lifesaving value of dedicated evacuation helicopters," he says, "it will be the end of aeromedical evacuation by medics."

His way of doing business is that the patient always comes first. Flying with him and for him, you don't look at the weather, you don't look at the enemy action, you don't look at the terrain. You focus on the patient, and you never leave the area without the patient. It may take two or three helicopters to get the patient out, but you get him out.

On July 1, 1964, Kelly is approaching a hot landing zone to pick up a patient when the enemy attacks. He's told to get out.

Kelly replies in his soft Georgia drawl. "When I have your wounded."

A round comes through the opened cargo door. It enters Kelly's right side, rips through his heart, exits, and lodges in the door on the opposite side. Kelly flinches. The blades hit the ground and the aircraft is pretty much destroyed. The physician on board breaks his legs, but the rest of the crew isn't hurt too bad, and they get Kelly out. He's dead.

A few minutes after Kelly is shot, I fly into the same area. My helicopter is shot up on the first approach, but we manage to rescue the patient.

I was supposed to replace Kelly on July 1. Today. I move my stuff into his room and spend the night sleeping in his bunk.

The battalion commander calls me in the next day and says, "The way you guys fly, I knew someone was going to get killed. But I didn't think it would be Kelly. I thought it would be one of the young pilots. I hope you're going to change your ways."

"We're not going to change anything. We're going to continue flying the way Kelly taught us because we don't know any other way."

As I'm leaving, he says, "Wait a minute."

The battalion commander hands me the bullet that killed Kelly. I still have it to this day. I've offered it to the family, but they didn't take it. They want the bullet in the National Medal of

AMERICAN HEROES

Honor Museum in Arlington, Texas, next to a display honoring Kelly and his DUSTOFF program.

I'm a pallbearer at his funeral. The outpouring is immense. Stillwell openly weeps. General Westmoreland singles out Kelly as a soldier for the ages. One life lost, hundreds of thousands saved.

Kelly took an American resource dedicated to American casualties and began using it for Vietnamese casualties. He spearheaded the humanitarian effort, unique in our past wars, in which US resources would be used for the benefit of the citizens of the country. We called it WHAM — "winning the hearts and minds" — essential in guerrilla warfare. The result was hospitals, orphanages, wells, vaccinations, and untold lives saved by the GI in the middle of the battle.

Kelly's dying words — "when I have your wounded" — set the standard for DUSTOFF to this day. He refined medical rescue resources to such an extraordinary degree that a soldier shot in a jungle in Vietnam had a greater chance of survival than if he suffered a car accident on a highway in America. He was one of, if not the most, extraordinary soldiers I've ever met or served with and, I believe, the greatest hero of the Vietnam War.

I return to Vietnam for my second tour in August 1967 with the 54th Medical Detachment. There are forty men, six brand-new Hueys, and twelve pilots, eight of which graduated the same day — one a teenager fresh out of high school. I have twenty-two hours to train them for combat, which will begin as soon as they get to Vietnam.

My first tour was around the Delta, a flat, large, forced landing area with endless canals facilitating navigation. Now I'm based in Chu Lai, in mountainous, tough terrain about twenty minutes south of DA Nang, a big city in the northern part of South Vietnam. I'll be dealing with jungles and mountains, no good emergency landing areas, tough navigation, and frequent, vicious tropical storms.

The worst part is the weather. In the mornings, we have a low valley fog, which comes up about five hundred feet and looks like a snowbank. Later in the day, we're dealing with the effects of cloud build-up on top of the mountains where the outposts are.

I'm scared to death of the weather and the mountains. These two factors, along with flying at night, are accidentally killing more of our Dustoff crews than the enemy. I know my new pilots are going to push themselves to retrieve a wounded comrade, but they're no match against Mother Nature.

I double my prayer routine.

The prayers work. I solve the night weather problem by using flares to rescue troops during a tropical storm in the mountains. We still face the day weather problem. One afternoon, I get a call that a young trooper has been bitten by a snake. He's stuck on an outpost atop a mountaintop blanketed by clouds.

I make several attempts but get disoriented and have to retreat to the valley. I have no idea how I'm going to get this kid out. I pray like crazy, asking the Good Lord to show me how to do it.

My crew is nervous and the troops on the ground are shouting that the snake-bitten soldier is going into convulsions.

"We got to give it one more try," I tell my crew.

Next time up, a gust of wind blows me sideways. I always have my right-side window open, and as I look out, totally disoriented and thinking I'm going to crash, I see the tip of my rotor blade and the tops of the trees.

That wind was the breath of God. Now I have two visual reference points—and I know I'm right side up. I turn the helicopter sideways and hover up the side of the mountain. We manage to get the kid and bring him to the hospital.

That's the flying technique we use going forward—low valley fog, afternoon build-up, two reference points, sidewards, and we can get in. The fog is perfect protection from the enemy. They can't see more than twenty feet in that stuff.

Now we're set. Nothing can stop us. Not night, not weather. When we get a call that a soldier has been shot or wounded, our aircraft is off the ground in two minutes. Picking the patient up and delivering him to the operating room of a Navy hospital, a hospital ship, or a surgical hospital—the average time is thirty-three minutes. Doesn't matter if the patient is a double amputee or has a sucking chest wound, he's going to live if we can get him there on time. In rough numbers, out of the ten percent of those who are wounded, less than one percent die.

When a mission comes in, the copilot and crew get the aircraft cranked and ready to go. Our helicopters aren't armed, but we—pilot, copilot, medic, and crew chief—are to protect ourselves and the patient.

We don't wait for helicopter gunships. It takes them too long to get off the ground, for one, and while they're extremely helpful to ground soldiers engaged in combat, they're virtually no use to us.

Nighttime is the safest time to fly. While night-vision goggles will go on to become a great military asset, we don't have them in Vietnam. When we do a nighttime approach, our console lights and outside lights are off—we're totally blacked out. Very seldom do we ever use a landing light or searchlight. They can expose the friendlies. The enemy, of course, can hear us coming, but they can't see us if the helicopter lights are off.

When I'm flying, I want it to be as quiet as possible so if someone starts shooting, I'll know it's the enemy, that I'm coming in the wrong way and need to come in a different way.

Using these techniques, and the Kelly spirit, the 54th rescues over twenty-one thousand patients in ten months.

On January 6, 1968, I'm woken up and told that some Vietnamese are at an outpost and weathered in from the fog. There have been seven attempts to rescue them. I'm not on duty, but I'm being asked because I've had success flying in bad weather conditions.

Using the flying technique from rescuing the snake bite victim, I go in and pick them up. On the way to the hospital, I hear about seventy-five or so other patients who are stuck in the low valley fog at LZ West.

"How long have they been there?" I ask.

"All night. Several aircraft have been shot down trying to rescue them."

I'm horrified that we haven't retrieved them. I drop off the patients and seek out the brigade commander.

"You can't get in there," he tells me, "and I'm not going to lift artillery."

I tell him we can. The brigade commander looks at my copilot and says, "Can you guys do that?"

"Yes, sir, we just did it."

"Okay. Given the number of patients, I'm going to line up four aircraft behind you. That way, you can show them how to get in there."

I don't like that idea. Once in the fog, I lose communications with them. Thank God they turn back. My crew will have to rescue the remaining patients. We go in and fly right over the enemy. We pick up the first group of patients and they're backhauled to the hospital. It takes four trips, but we get everyone out.

On the way back to LZ West, we get a call from Savage Golf.

"I've got two urgent patients," he says. "The area is secure. I'll mark it with purple smoke."

I make a low, fast approach, and as we hover over the smoke, I don't see anyone.

"They're hiding in the grass," my crew says as we come under fire.

The helicopter takes several rounds. We jump out of there and go up to altitude to see if we're still flyable. The instruments are good, and the aircraft seems to be flyable.

Savage Golf comes back on the air, begging us to return. I decide to have a "come to Jesus" talk. "Give us the enemy location and get off your ass and we'll come in again. And hold your fire. I want a quiet approach. If I hear fire, I'll know it's the enemy and abort."

They stand up and help load the patients. We take fire but don't get hit. When we arrive back home, we learn that the controls have been partially shot away. I get another aircraft, and then we go out again.

We're monitoring traffic with a Dustoff aircraft from a sister unit that is on the ground in a minefield. Before the Dustoff could get the patients, a mine exploded near the aircraft, killing two more soldiers. I can hear the men on the ground screaming over the radio. The damaged helicopter is forced to leave the area.

I saw where the helicopter was sitting. If I can hit that spot or damn close to it, I probably won't set off a mine.

Two problems. I could land on another mine — or the mines down there might be command detonated. Second, if there's a power change from the down blast of the rotor blades, I could set off a mine.

I go in and pretty much land on the previous helicopter's skid marks.

No one outside will move. I turn to my medic and crew chief and say, "Go get them."

I watch through the cargo door as they jump out and start dragging people through the minefield and back to the helicopter. Nobody will help them, nobody will move.

My medic and crew chief are carrying one patient on a litter when a mine explodes. It blows them up in the air. *They're going to go through the rotor blade,* I think as the aircraft fills with shrapnel.

They land...and get back up. They're both hurt, but not seriously. The patient on the litter, I'm guessing, took on most of the mine's shrapnel.

We deliver the wounded and get a new helicopter. My medic and crew chief are the real heroes and are awarded Silver Stars for their actions.

In the fall of 1969, I'm an instructor at Fort Sam Houston, Texas, when I get a phone call.

"Major Brady, this is Major Scott from General Westmoreland's office. Congratulations. You will receive the Medal of Honor for actions on 6 January 1968 in Vietnam."

This has to be a practical joke — which wasn't uncommon among my fellow troops at the time. More puzzling is the fact that I've just been awarded my second Distinguished Service Cross for that action.

"Are you sure?" I ask.

Major Scott assures me the Distinguished Service Cross is an interim award while they finish processing the Medal of Honor. I'm completely surprised. And the beautiful thing is that I get to bring my family and a lot of my friends to the White House on October 9.

Before the ceremony, President Nixon says, "Did you know the Medal of Honor Society is meeting at the Shamrock Hotel in Houston, Texas?"

I have no idea what the Medal of Honor Society is. Neither do the three other soldiers who are being awarded the medal with me.

"Those are the living recipients of the Medal of Honor," President Nixon explains. "And you will be one of them this afternoon. Would you like to go?"

"How do we get there?"

"You can take Air Force One."

The ceremony is held outside on the White House Lawn in front of five thousand people. As I'm standing next to President Nixon, I look out at the crowd, at the pilots and crew who did all I did and more, and, frankly, feel a little bit embarrassed. Every other pilot in my unit, except for the weather missions, did the same thing I did, got shot down as many times as I did, carried as many patients as I did (or almost, as I had a year of experience on them). They should be getting the same recognition.

We fly to Texas, four young troopers, on Air Force One. When we enter the ballroom of the Shamrock Hotel, we're given a standing ovation from several hundred Medal of Honor recipients going back to the Boxer Rebellion. Military legends like Jimmy Doolittle, Eddie Rickenbacker, and Joe Foss. Bob Hope and Dinah Shore are the night's entertainment.

It's an amazing experience.

I was a reluctant soldier, God knows. I was forced to join the military, didn't want anything to do with the uniform or anything about it. Had I not gone to Berlin, had I not had to do things I didn't want to do and would have never done if it was up to me—the lesson is, I wouldn't trade a minute of it. It was the greatest thing that ever happened to me. I was able to help rescue over five thousand wounded soldiers, an experience beyond all others.

AMERICAN HEROES

I think that young people today need to have the same opportunity to serve their country, to do something for somebody else besides themselves, which is what military service is all about. Soldiers believe life has no meaning unless it's lived for the benefit of future generations. We don't believe we did America a favor by our service and sacrifice, we believe God did us a favor by allowing us to be born in this great country. That's what it's all about. That's why they protect America. It's not just for them, it's for future generations.

Paris D. Davis

Captain, US Army
Conflict/Era: Vietnam War
Action Date: June 17–18, 1965
Medal of Honor

Not long after I entered the Army, I had just completed the IOBLC, the Infantry Officers Basic Leadership Course, when a sergeant major who is white comes up to me. "What are you doing here?" he asks.

It's been fourteen years since President Truman desegregated our military, but Black people are still looked upon as less than people — less than Americans. Sitting at a lunch counter, getting a book from the library, walking a picket line to support the right to vote and integrate schools and public transportation — these actions can get Black people arrested, beaten, or killed.

I straighten a bit. "I'm waiting to be assigned, sir."

He looks at me, thinking.

"I have an Airborne slot," he says. "You want it?"

"Yes, sir."

"Do you know what 'Airborne' means?"

"No, sir."

"Good. You'll find out once you get there."

I start training—running, push-ups, everything—and go to Airborne School.

As the airplane levels off for our first jump, everyone is crossing their fingers. Some guys wet their pants. When I get to the door, I don't want to jump. I have to be pushed out.

I manage five jumps without breaking my legs.

The sergeant major sees me and my new jump wings and says, "What's going on now?"

"I'm still waiting to be assigned, sir."

"I've got an assignment for you. You're going to work for me for a while, and then I'm going to send you...let's call it a surprise."

The "surprise" is an invitation to the Ranger Course.

Rangers are predominantly white. People pull me aside and say, "Are you sure you want to join? There aren't a lot of people like you in this outfit."

I don't listen to them, and enroll. Two other Black soldiers are selected to go through the Ranger Course. They're in the same company as me. On the second day, they run into the hooch, pack all their stuff into a Jeep, and drive away.

I'm assigned a "buddy"—a blond guy who is six foot two and built like he whips butts all day long. During judo class, he throws me around like I'm a rag doll. Breaks my lip several times and bangs me up, but I hang in there with him. He takes a liking to me, and we become friends.

Together, we go through Ranger School together without

any problems. The experience teaches me the importance of having a buddy.

I'm now an Airborne Ranger, After I graduate, I'm sent to South Korea. While I'm there, I hear about President Kennedy expanding Special Forces. I decide I am going to volunteer, so that's what I do. I'm determined to become a Special Forces soldier. The distinctive Green Beret is, as President Kennedy says, "a symbol of excellence, a badge of courage, a mark of distinction in the fight for freedom."

After I complete my Special Forces training, I'm one of America's first Black Special Forces officers. A Green Beret. I get my next assignment. I'm going to Vietnam.

Back home, the United States is engaged in another kind of war. It's a battle over segregation. In 1964, President Kennedy's successor, Lyndon B. Johnson, signed into law the Civil Rights Act, which ended segregation in public places and prohibited discrimination on the basis of race, color, religion, sex, or national origin. Not everyone is happy about it. Some white people cross to the other side of the street when they see me. Some white soldiers I know, too.

Protests are erupting across the country. On March 7, Dr. Martin Luther King, Jr., and a group of nearly six hundred people march from Selma, Alabama, to the state capital in Montgomery, in a peaceful protest about Black Americans being denied the right to vote. They reach the Edmund Pettis Bridge when Alabama state troopers, armed with tear gas, nightsticks, and whips, beat the protesters back to Selma.

Less than a month later, I'm back in Vietnam for my second

mission. I've been promoted to captain, and I'll be serving as the commander of Alpha Detachment A-321, 5th Special Forces Group (Airborne), 1st Special Forces.

And because I'm Black, my commander warns me, I'm going to have to work twice as hard.

"You're going to have an all-white team, and you have some guys from Alabama and one from Mississippi — it could be a rough thing," he says.

I address my men. "If you call me anything besides 'sir,' I'm not going to waste time reporting you, I'll just knock you to the ground."

We get along splendidly. I think one of the good things about a war or any type of crisis like Vietnam is the fact that people that are committed to it gel. There's no race here. In the dark, brown is just as black or white as anyone else.

One of my tasks is training South Vietnamese volunteers in Binh Dinh province to serve in the Popular Forces. The "Ruff Puffs," as they're called, dress like the Vietcong, in shorts or loose black clothing, and patrol at night, armed with American M16 rifles. They are exceptionally skilled at ambushing the enemy.

What I do know is that when it comes to soldiers, if we don't work together, we won't get the job done. If I get with a group that ostracizes me because I'm Black and they're white, when something comes up where I need their help, they won't do anything, and I'll be left by myself.

My men and I build a camp together in Bong Son. The camp is an outpost in hostile territory. By necessity, we become a tightly knit group.

I get some schooling on how to fight the Vietnamese from William "Billy" Waugh, a fellow Green Beret who will later be recognized as one of the best American jungle fighters who ever lived. He teaches me the importance of cross-training soldiers. On patrols, if you can't have a medic with you, then you want, say, a weapons guy who has been cross-trained as a medic.

The second thing he teaches me is the importance of protecting your camp.

We pay the locals to help with the building. During those first few weeks, our senior medic, Hugh Hubbard, makes it a point to get to know the families—especially their babies and children. He examines their scars and runny noses and other maladies and always offers some medical solution.

Hubbard gives the kids little tubes of toothpaste. They think it's candy. They chew it and blow bubbles. The kids are fascinated by our monkey. His name is Joe, and Hubbard has taught him how to read colors. While the kids sit in the clinic, coloring with crayons, Hubbard will tell Joe to grab some box based on its color. Joe climbs a pole and always comes back with the right stuff.

The families are grateful, and now we have an added layer of protection. The Vietcong see we have so many of the indigenes working for us, see how well we're treating them, and decide to leave the camp alone.

When it comes to fighting, we have a pretty good record. Our actions gain the attention of General Westmoreland, the commander of US forces in Vietnam.

"We have two other divisions here," he tells me. "Why are you doing better than both of those divisions?"

"Sir, I'm not going to get into a battle unless we have an advantage. If we don't have an advantage, we can't win."

On the evening of June 17, I lead three Green Berets and about 100 Ruff Puffs on a ten-mile march to conduct a raid northeast of Bong Son. The raid lasts well into the morning. We capture four enemy combatants who, during questioning, reveal the location of a well-trained and well-armed Vietcong headquarters equipped with two hundred to three hundred enemy soldiers.

Ron Deis, a junior member of my Special Forces team, is our spotter. He's flying in an L-19 Bird Dog propeller plane. I relay the information to him. As the sun starts to rise, Deis confirms the enemy-obtained intel is correct.

I set off for the camp with Sergeant Major Billy Waugh; my medic, Spec-4 Robert Brown; and Staff Sergeant David Morgan, my demolitions specialist. Accompanying us are roughly seventy Ruff Puffs.

The Vietcong general, we've been told, sleeps in a big tent in the middle of the compound. We find it easily, and when we enter, a woman jumps straight up from the bed. She's bucknaked—and armed. She's getting ready to take a shot at me when Waugh shoots her.

The general is also here. He's turning his weapon to Waugh when I shoot him.

The battle is on.

We go from hooch to hooch, throwing grenades and engaging in firefights. Some NVA soldiers fight back. Others flee.

There are several times when I'm forced to engage in hand-to-hand combat. Sometimes I use the butt of my rifle as a weapon.

I hear a bugle. The sure sign of a counterattack.

We push forward, heading to what we believe is the enemy's command building. I open fire with my M16, and when I reach the building, I throw a grenade through a window. Then I move inside and open fire again, taking down several NVA soldiers. I sustain a wound to my forearm.

I radio for more ammo as hundreds of enemy combatants begin to converge on our position. Within minutes, the jungle lights up like it's the Fourth of July. The blast from a grenade takes part of my trigger finger and several teeth. The spotter aircraft with Deis is damaged in the fighting and flies back to base camp.

I spot a South Vietnamese commander fleeing the ambush. I leave the rice paddy and chase after him. He escapes. By the time I return to the rice field, a sniper has put five shots in Sergeant Major Waugh's foot and leg. He's pinned down in the paddy, inside a buffalo rub, a big hole where oxen roll around on the ground.

My medic, Brown, has been shot in the head. An exploding mortar has taken down my demolitions specialist, Staff Sergeant Morgan. He's stuck in a cesspit on the far side of the field, and he's taking fire from the enemy sniper.

The enemy is coming at me in waves. I manage to move myself and the rest of my soldiers to a hill where the NVA has left behind a dozen or so foxholes. I need to call in for artillery and air strikes. I use my PRC-10 radio but communications are down.

The Vietcong tries to overrun us. I see five coming over the trench line. I kill all five when I hear firing from the left flank. I run down there and see about six Vietcong moving toward our position. I throw a grenade and kill four of them.

My M16 jams, so I shoot one with my pistol and hit the other with the butt of my M16 again and again until he's dead.

That's when it hits me.

I'm the last American standing.

Sergeant Major Waugh is crying for help.

"I'm coming for you!" I yell. *"I'm coming for you!"*

Then I hear Morgan yelling from a ditch.

I search for the sniper. Finally, I find him and his camouflaged hiding site. I kill him with my rifle and then crawl to his site and lob a grenade in it.

Two more Vietcong are dead.

Morgan is stuck in a trough of human shit the farmers use as fertilizer. I throw him a rope and pull him free.

The Ruff Puffs are young, and they don't have any fighting experience. They're staying in the fight, but some, I believe, are breaking under pressure. Their numbers are dwindling, and the enemy is coming at us from all directions. Bodies are stacked on top of one another, creating walls of human flesh; I'm beginning to run dangerously low on ammo, and I'm worried that we're going to be overrun by the enemy.

I pick up the radio and manage to regain communication. A medevac helicopter, I'm told, is inbound, just a few minutes away.

I call for an artillery strike within thirty meters of my position. Any artillery strike within five hundred meters of friendly position is considered "danger close," but I need to take the risk. I need to kill as much of the enemy and push the remaining ones back if I'm going to have a chance of rescuing Waugh.

After the Army rains down hellfire, I sprint from my position—and immediately take on gunfire. I'm shot in the arm. I turn back and treat the wound the best I can, with what I've got, and then I wait for another window of opportunity.

The Army fires more artillery. I sprint again from my position, and this time I make it to Waugh.

Enemy fire forces me to retreat and find cover. I'm going to get Waugh out of that buffalo rub. A Green Beret doesn't leave anyone behind. It's our mantra. The code we live by.

Again, I sprint to Waugh. The enemy gunfire is intense.

Again, I can't break him out of the buffalo rub.

And now, the enemy is about to swarm on our position.

I pick up a machine gun and start firing. I see four or five of the enemy drop. The remaining ones break and run. I then set up the 60 mm mortar, drop five or six mortars down the tube, and then run back to Waugh.

This time, I have help. A fellow Green Beret, Sergeant First Class John Reinburg, joins the fight. He arrived here on the medevac helicopter, and it's waiting at the top of the hill. Reinburg begins running ammo to the remaining Ruff Puffs.

Waugh can't walk. I drag him away and then pick him up and carry him. I'm halfway up the hill when I'm shot in the leg. I keep walking and deliver Waugh to the waiting helicopter.

I'm with Reinburg when he takes two rounds in the chest. I lift his 240-pound body in a fireman-carry—four hundred meters across the muddy rice field to safety.

My thoughts never leave my medic and fellow Green Beret, David Brown. Brown is a brand-spanking new dad. The day before, his wife gave birth to their first child. I have no idea if he's dead or alive, but he's somewhere out there, and I'm going to find him.

A rescue helicopter has landed. Major Bill Cole, my commander, is here.

"Leave with the wounded," he tells me. "I'll relieve you."

I've been fighting for roughly ten hours. "Sir, please don't do that to me. I'm not hurt that bad. I still have an American out there."

"You've got it, Davis. Good luck and God bless you."

Late in the afternoon, a badly wounded Vietnamese interpreter tells me he found David Brown.

"He alive," he says.

Air support is circling overhead. I'm on the radio, directing tactical air and artillery fire on the enemy's position, when the Air Force colonel who is flying high above the battlefield and serving as a forward air controller orders me to leave the area.

"That's not going to work, sir."

He insists that I leave. I insist that I'm not going to leave until all my men have been recovered.

I'm disobeying a direct order.

After we have a heated exchange, he says, "I wish I was down there."

There's plenty of room, I say to myself.

The colonel doesn't join the fight, which is probably a good thing. If I saw him, I'd probably kick his ass.

As fighters drop bombs, I begin a 150-yard crawl through mud and human waste. Enemy forces fire at me and lob grenades. Shrapnel singes and pierces my skin. Finally, I reach Brown.

My medic is drifting in and out of consciousness, his body covered in leeches. His head is bandaged. Brown tells me a Vietnamese medic did it, before the man was killed. Brown says he was shot in the head early in the battle and has been lying right here, in this rice paddy, ever since.

That was over fourteen hours ago.

"Am I going to die?" Brown asks. "Am I going to die?"

"Not before me."

Shortly after the battle, Major Cole nominates me for the Medal of Honor. Billy Waugh does, too. He personally writes a letter while he's receiving medical treatment for his injuries at 5th Special Headquarters in Nha Trang, Vietnam.

"I only have to close my eyes to vividly recall the gallantry," Waugh writes. Major Cole tells a newspaper that I "showed as much cold courage as any human I've ever heard of."

Later that year, I'm awarded the Silver Star, the Bronze Star, the Air Medal, and the Purple Heart. Years pass, and I don't hear anything about the Medal of Honor.

In 1969, I'm told the Army can't find the nomination package.

A hearing is held. Major Cole states he submitted the

nomination packet to the Special Forces headquarters. From there, it should have gone to the Pentagon, but there's no evidence it arrived—or was ever sent.

The conclusion is that the packet was either lost or destroyed. Major Cole is ordered to create and submit a new packet.

It disappears. Again.

A Medal of Honor packet requires substantial paperwork—eyewitness statements, map, a unit report of the action, and other documentation. It's not something that gets lost. The chances of it being lost not once but twice are nearly impossible.

Race, I believe, is a factor.

"What other assumption can you make?" says Ron Deis, my fellow Green Beret who was on my team in Bong Son and flew over the battlefield that day in Vietnam.

When you're out there fighting, everybody's your friend, and you're everybody's friend. The bullets have no color, no names.

I think about a pilot I rescued while on another mission. He was white. I was at Fort Bragg and saw him with his wife and child, and when he saw me, he crossed to the other side of the street. If I had been white, he would have come over and hugged me.

In 1981, with still no word about my nomination, Waugh writes another personal statement. Waugh, who was awarded the Silver Star and has gone on to have a legendary military career, is told that my packet is, in fact, working its way through the system.

"My mind is fixed on it," Waugh says. "Davis did a good job, and I am proud of him."

The new recommendation again goes nowhere.

I retire from the Army in 1985. I start a small newspaper in Virginia called *The Metro Herald* and, for the next thirty years, publish stories about the accomplishments of Black people. I'm often asked about my thoughts on why my medal nomination kept getting lost or why I kept running into enemy fire to save my men. I always answer the same way.

"Life suddens upon you, it just suddens upon you. Every day, something comes up that you don't expect."

In 2014, a diverse, fifteen-member group of veterans and volunteers with legal, research, and communications expertise resume efforts to revive my Medal of Honor nomination. They assemble an updated awards package comprised of Army files, after-action reports, new and old affidavits and interviews, long-lost news reports. They soon get the package in the hands of a new generation of Pentagon decision-makers.

Behind the scenes, the volunteer group continues to meet and make hundreds of calls and presentations to military, political and veteran leaders. This goes on for several years. I am unaware of most of the details of their work during this time. Then, in mid-2021, they introduce me to *CBS Mornings* and the *New York Times*, which first share my story with the American public.

For the next two years, CBS's ongoing reporting and our team help keep the story in front of high-level decision-makers. As their efforts go on, they begin to get quiet, encouraging signs from the Pentagon. Then, in early 2023, more than fifty-seven years after that battle in Vietnam, I get a call at home from

President Biden. He tells me I will be awarded the Medal of Honor and to prepare for a White House ceremony.

Speaking with the president prompts a wave of memories of the men and women I served with in Vietnam—from the members of the 5th Special Forces Group and other US military units to the doctors and nurses who cared for our wounded. I remain so very grateful for the support of my family and friends within the military and outside it. Their work, the White House ceremony, and many events at the Pentagon and elsewhere in America keep alive the story of A-team, A-321 at Camp Bong Son.

Most of all, I want to share the medal with my Special Forces troops—the other soldiers I worked with and fought with that day. Somehow, they need to touch that medal. It ain't all mine. It's for America, too.

Thomas William Bennett

Corporal, US Army
Conflict/Era: Vietnam War
Action Date: February 9–11, 1969
Medal of Honor

Interview with George Bennett, brother

Someone is banging on the window in the upstairs dormer.

It's Tommy, my youngest brother. He's out after curfew. I'm usually up late studying for my college classes, so he jumps from the front porch onto the roof, then keeps tapping on the glass until I let him in.

Like me, Tommy works very hard to live up to expectations. He is friendly, outgoing, genuinely interested in people, and was voted "Most Polite" his senior year in high school. Tommy is not out drinking and carousing and causing trouble. He's breaking curfew to attend various religious group sessions.

Our family is Northern Baptist. Tommy is very serious about

spirituality and wants to explore all religions. He thinks our mom and stepdad will be offended by his late-night activities, which is why he's keeping them a secret.

A less well-kept secret is that Tommy pushes himself to the limit. In high school, the principal was afraid Tommy was going to make himself sick and made him step down from his position as the president of the student body.

The loss *crushed* my brother. But he's accepted it. And Tommy has other more important things on his mind, like Vietnam. *Especially* Vietnam. People are burning their draft cards and fleeing to Canada. Friends and counselors are urging Tommy to do the same.

Tommy, though, is intensely patriotic. Our biological father passed away when I was eight and Tommy was five, and we've been raised by our stepfather, Kermit, a wonderful patriot who served in the Navy during World War II.

"Don't give up on your country," Kermit always tells us.

A lot of Tommy's friends have already entered service. His good friend Dave Kovac joined the Marines and was killed in action. His death left a huge imprint on my brother. Running away to Canada, Tommy feels, would be disrespectful to Kovac's sacrifice for our country. Tommy has told me that if he's drafted, if he's forced to kill, he doesn't think he can remain faithful to his religious beliefs.

Choosing between two loyalties—his country and his God—is torturing my little brother.

Fortunately, he doesn't have to worry about the draft. Tommy will be attending West Virginia University in the fall. He'll earn a student deferment and most likely miss the Vietnam War.

* * *

On college campuses across the country, students are protesting. They organize national burn-your-draft-card movements against the Vietnam War. They demonstrate against Presidents Johnson and Nixon, the draft, the Pentagon, and Dow Chemical, the maker of napalm. At Harvard, Secretary of Defense Robert McNamara is trapped inside a police car while protesters shout questions at him.

At West Virginia University, Tommy is as involved in student government organizations as he was at Morgantown High. He's a well-known campus peacemaker and a skilled diplomat, equally respected by students and administrators. By bringing people together to talk things out, he helps keep peace on campus.

As president of the Ecumenical Council, he writes the student code of ethics and leads Bible studies. He teaches Sunday school and ministers to local customers on his bread delivery route.

My brother is very serious about life. "Who are we as a people?" he asks himself, and "Isn't there some way for the world to get along better than it currently is?"

In pursuit of answers, his grades suffer, and at the end of 1967's fall term, Tommy is placed on academic probation. If he loses his student deferment, he'll become eligible for the draft. If he's drafted, he can serve or leave the country — or he can declare himself a conscientious objector.

Tommy learns about another option. He can apply as a conscientious objector who is willing to serve. That was the path taken by Desmond Doss, a Seventh-day Adventist who served as an Army field medic during World War I. Though Doss didn't believe in violence and refused to touch, let alone train with, firearms, he rescued seventy-five men in a grueling battle and was later awarded the Medal of Honor.

Tommy deeply loves his country. Running away to Canada is not an option.

Tommy decides to enlist as a conscientious objector. His request is granted on May 2, 1968. He won't have to handle a weapon when he goes off to Fort Sam Houston for medic training.

I can't believe my brother is going to be a medic. Tommy is very skittish about blood. He literally passes out when having his blood drawn.

He writes a letter to my parents saying there's a strong chance he won't be on the front lines because the number of medics going to Vietnam has decreased significantly. "I have just as good of a chance to serve in Honolulu as in Nam," he writes. "Even if I go to Nam, I might not serve in a combat zone."

PEACE is written in capital letters on the bottom of the page.

Six weeks later, another letter arrives. "If I am called to Nam, I will go," Tommy writes. "Out of obligation to a country I love I will go and possibly die for a cause I vehemently disagree with."

Two weeks before graduation, Tommy is told his entire medic trainee class is going to Vietnam.

Tommy returns home for the Christmas holidays in good spirits. He picks out an ugly tree for the family like Charlie Brown did in *A Charlie Brown Christmas*. He will deploy after the first of the year.

One night at dinner he breaks down.

"I just can't do it." He sobs. "I can't go over there. Mother, I'm too young to die."

We do our best to console him. My brother takes solace in a prayer he's written.

Oh, God, shake me from my apathy.
From my wanderings of mind.
Create in me discipline, concern, and love.
Help me live: really live: in Spirit and the Truth.

We celebrate the New Year at my aunt Mary's home in Fairmont, West Virginia. On January 5, 1969, we drive Tommy to the Pittsburgh airport. It's an emotional send-off, for all of us.

The only way Tommy can communicate with us is through the mail. He brought a reel-to-reel recorder with him and sends us tapes. He writes a letter or cuts a tape every couple of days.

Tommy has been assigned to Company B and will be flying out to meet his new unit, 1st Battalion, 14th Infantry. "Tomorrow, my journey begins," he writes.

Yet his patriotism never wavers. "I'll stick with her," he writes of America.

During dark, sleepless Vietnam nights, Tommy sometimes hears explosions and serves on "listening patrols." He writes, "It's hard to believe there are people out in those woods who want to kill me."

In early January, Tommy writes home that he likes being

called "Doc." He treats cuts and sunburns. One soldier gets bitten by a rat. The men in his unit are his new friends.

His unit treks up Chu Pa Mountain, a dense jungle full of ravines. "Several times we were climbing straight up," he writes. "My shoulders were full of pain."

The gear they're humping is so heavy, one man pulls a groin. Everyone helps carry the injured soldier and his equipment up the mountain. Tommy gets choked up.

"War is a terrible thing. I HATE it," my brother writes. "Yet it seems to bring out the best [in] these men. Now that we've made it to the top, there is very little danger."

On January 24, Tommy waxes philosophical. "I'm learning more, faster probably than I've ever learned," he says. "And, of course, the more you learn, the more you see, and the less concrete everything is in your mind."

On the tape, his voice sounds more at ease, with himself—with everything. "First of all and foremost," he says, "I am still safe. Second, I am ready for death, all the way around, and I'm proud that I'm ready. Third of all, and most important perhaps, how much I really love you."

It's February when two military men appear at my childhood home to tell my parents that Tommy is missing in action.

My stepfather has suffered two heart attacks. Soon after learning the news about his stepson, he has a third.

I'm a student at Carnegie Mellon in Pittsburgh. I call the military and get the number of one of the men who visited my parents. He picks up the phone.

"If you have any other bad news," I explain, "I want you to call me in Pittsburgh so I can come down and be with my parents."

He says, "I can't do that, George, because the parents have to be the first to know if the news is bad."

"You can have them be the first to know, but I want you to call me and say, 'It's raining in Morgantown.' That's all you have to say. After that, I'll get in my car, and I'll be back in Morgantown in an hour and a half."

I soon get a call that it's raining in Morgantown.

I drive home, thinking about the last tape from my brother—the one dated February 5—where he downplayed the danger he was facing.

"When you start adding up figures and taking percentages and stuff," he said, "over here there are very few places that I can be safer than with the US Army."

I recall more lines. "I have had such a rich, full, good, exciting life that, well, nobody can take that away from me." He said, "It can't be erased or diminished in any way from me. There's very little chance that anything's going to happen. And if it does, so what? I've had my twenty-one good years."

I arrive home. I'm literally standing at the front door when the military arrives to tell us that Tommy, my twenty-one-year-old baby brother, has been killed in action.

The day of Tommy's funeral, I step out of the church to see a line of cars in every direction. A lot of those people, I learned, were the ones he ministered to on his bread route.

* * *

My mother picks up the phone. The person on the other end of the line says her son has been nominated for the Medal of Honor. The award will be presented on April 7, 1970—Tom's birthday. Had he lived, my brother would now be twenty-three.

His death was a life-changing event for all of us, but more so for my mother. She's never been the same. My mother doesn't want to accept the medal—or even go to Washington. After Tommy died, she wrote to President Nixon, explaining her son's belief in nonviolence. Her absence from the ceremony, she decides, will be her protest against the war and Tommy's senseless loss.

My stepfather tries to convince her to reconsider. "It was the boys in his outfit that put him up for it," he says. "They wanted him to have it."

It takes some time for her to agree.

At the ceremony, my parents and I meet the soldiers who petitioned for Tommy to receive the medal. From them, and from Tommy's citation, I manage to re-create my brother's final days.

On February 9, Bravo Company is moving downhill when they hear intense fire coming from an AK-47. Bravo's sister unit, Delta, has walked into an ambush. Tommy's platoon is ordered to hit the enemy from the rear.

They don't get far until they, too, are ambushed.

Three men are down. Everyone dives for cover. Everyone except Tommy. He crawls his way to each man and renders lifesaving first aid. After assisting each of the three men, Tommy, all of five-feet and six-inches, carries each one to safety.

The firefight isn't over. The enemy is well-fortified and outnumbers the American soldiers. Anytime someone cries out for help, Tommy makes his way across the battlefield, dodging fire from small arms, automatic weapons, mortars, and rockets.

Finally, the enemy pulls back. Medevac choppers arrive just before nightfall. Five dead soldiers and six others who are seriously wounded are removed from the battlefield. Instead of sleeping, Tommy spends most of the night tending to the wounded soldiers who haven't been evacuated.

James McBee is Tommy's platoon sergeant. McBee approaches the captain and says, "Sir, the men have asked me to put Corporal Thomas Bennett in for a Silver Star. He's been doing an outstanding job today. He took a lot of risks to help the guys who got hit. In fact, I had to kind of chew him out for taking too many risks."

"What'd he say?"

"He said he wasn't afraid. Said he was trained to be a medic and he was just doing his job. Said the Lord would protect him and if he died it would be God's will."

"Okay," the captain says, "I'll write him up."

The next day, Bravo Company is attacked again, this time by B-20 rockets. They take out clusters of men. The fire from AK-47s is intense. Tommy ignores the danger and makes his way to each wounded man, giving morphine and slapping on bandages.

"Don't worry," he tells each soldier, "you'll be okay."

Tommy spends another sleepless night tending to the wounded.

On February 11, as the rising sun parts the darkness, Bravo

Company is targeted by snipers. A lot of soldiers are taken down, and through it all, Tommy risks his life so he can be by their side.

"Be careful," Sergeant McBee warns him.

Another soldier — a young private new to the platoon — is hit by a sniper round. He's roughly thirty feet from Tommy and crying out for help. Tommy is about to move to the man when McBee grabs him and says, "Don't go out there! He's gone."

Tommy ignores McBee and jumps up, intent on saving another life.

A bullet hits Tommy below the rim of his helmet. More bullets rip through his body as he falls, but it doesn't matter. Thomas William Bennett, brother, son, Army medic and conscientious objector, a man of deep faith and conviction — everything Tommy is or will ever be — is already gone.

The three men my brother carried to safety on February 9, 1969, are at the ceremony. Each one seeks me out, looks me right in the eye, says, "Your brother saved my life."

The feeling is indescribable.

It's been fifty-five years since my brother died. My mother and stepfather are now buried by his side.

Yet his name lives on. Tommy's had medical facilities named after him. A bridge. A youth center at Schofield Barracks on Oahu, Hawaii. A dormitory at West Virginia University is named Bennett Tower. The ecumenical house on Oakland Street in Morgantown is called Bennett House, and his name is written on a plaque with other veterans who were students at Morgantown High School and died in service.

A photo of my brother hangs in the Hall of Heroes at the Pentagon, surrounded by those of others awarded the Medal of Honor.

What amazes me — what continues to amaze me — are the people Tommy never got a chance to meet. The people who send me pictures of them leaving flowers at his gravesite, especially on Veterans Day. Bonni McKeown wrote a book on my brother called *Peaceful Patriot: The Story of Tom Bennett*. Many students at our former high school read the book as part of the English curriculum.

Tommy touched the lives of so many people beyond his family. My baby brother packed a lot of life into his twenty-one years.

Cindy Pritchett

Command Sergeant Major General, US Army
2002 Living Legend Award, Army Center Heritage Foundation

The best way to piss off a Navy man, I figure, is to join the Army, so I enlist. I'm eighteen and don't need any parental consent.

We live in Mount Clemens, Michigan, outside of Detroit. It's July of 1973, and the city is plagued with race riots. The Vietnam War is winding down, so I'm not concerned about being sent overseas.

I don't tell my father. We don't really get along. Our latest conflict is over where I'll go to college. I want to go to a local state university. My father is very impressed with my cousin, who is going to Notre Dame. He's always trying to get me to keep up with my cousins.

Two weeks before I ship out for basic training, we get into another big fight.

"Guess what, Pop? I've joined the Army."

He stares at me in disbelief.

"You don't have the guts," he says.

I throw my enlistment papers at him. *"Read it and weep."*

I join the Army, I won't have any of Daddy's friends keeping an eye on me. Both my parents were in the Navy— my mother as a Navy Storekeeper, my father in aircraft maintenance. He served for twenty-five years before retiring.

When I arrive at Fort McClellan, Alabama, I call my mom.

"There's somebody here who wants to talk to you," she says.

My dad comes on the line. "Look," he says, "you can go to whatever college you want, but just come on home."

"Pop, you of all people should know I can't come home now."

I enlist at the start of the all-volunteer Army. That doesn't mean it's modern. The Women's Army Corps (WAC) created during World War II is still in existence. The WAC Center, where we train, was established at Fort McClellan in the 1950s.

Our uniforms are white tennis shoes, black bobby socks, a blue shirt, blue shorts, and a wraparound skirt that is removed when we do PT. We learn how to apply makeup, take classes to correct our posture, and learn to walk balancing a book on our heads — class after class after class reinforcing us on how to be a lady. We're shown weapons and what they can do, but we are *not* allowed to fire them.

I graduate from basic training and get sent to Fort Knox, Kentucky. I'm assigned to the WAC detachment, where I learn my craft as a supply specialist. After my time at Fort Knox, I'm assigned back to Fort McClellan, to the same unit where I was a trainee. I'm now the unit supply clerk. After about a year, I volunteer to attend drill sergeant school. I graduate number one in my female class and get a merit promotion to sergeant. All the drill sergeants in WAC basic training at the time are female. I become a drill sergeant—and love it.

The Army starts bringing in male drill sergeants for female basic training. A male senior drill sergeant named Sergeant First Class Odin pulls me aside and says, "I don't know how to train women."

"Tell you what," I reply. "I'll teach you how to deal with women if you teach me everything you know about infantry." Women are finally allowed to learn how to shoot weapons and do more field training.

Odin agrees.

I say, "Do you have daughters?"

"I do."

"Training women is no different than dealing with your daughters. There are plenty of times when you tell them no, right? And when you say no, they try to wrap you around their little fingers. Some female basic trainees are going to try to manipulate you in the same way, so you need to be on guard for that."

Then I speak to him about "after hours." Every night, a drill sergeant is on duty for what's called "charge of quarters duty," or CQ, which includes a bed check.

"As a male, never do a bed check alone," I tell him. "You either make sure one of the female CQs is with you or someone else you trust because there will be that one woman who will challenge you. They'll use their wily ways. And they *will* cry or try to get you in a compromising situation. It's all about how you react to them."

One night, a male drill sergeant who's scheduled for CQ tells me his wife has gone into labor. I tell him to go to her and that I'll take over his CQ duty. I walk into one bay to conduct bed check and find a young woman out of bed, standing in the Aisle wearing her bathrobe.

I don't know why she's out of bed. All of a sudden, she opens the top of her robe a bit. I don't think she realized who was conducting bed check tonight, and I say, "Does nothing for me. Put your clothes on and go downstairs to the CQ desk. I'll deal with you later."

Drill Sergeant Odin is as good as his word. He teaches me how to be a soldier in this man's Army, the infantry, fields of fire, how to dig a fight position and how to assemble and disassemble the M16 rifle and the M60 machine gun. When women are finally allowed to learn how to fire the M16, I share the knowledge DS Odin taught me with my trainees.

I do a three-and-a-half-year tour at Fort McClellan.

My next duty assignment comes in 1978, the year the WAC is discontinued. I go to Germany, where men and women live in the same barracks. I'm the first female NCO for the 37th Transportation Group. I'll be responsible for supply, logistics, transportation, and maintenance.

The young soldiers aren't quite sure how to deal with me. They make a point of saying, "You're not a drill sergeant anymore," to which I want to say, "Yeah, I got that. No problem."

Some of the young guys, especially the ones who've never had a female supervisor, try to challenge my authority.

Fortunately, the master sergeant has my back, telling them, "She's the supervisor of the supply room. You guys are going to listen."

We all need people, male or female, who recognize some quality in us that's worth nurturing. People who can guide us, who will be truthful about where we're at, where we fall short, and what we should be doing to be successful.

A lot of my early success is due to male mentorship, but the reverse has also happened.

When I become a sergeant first class in the early nineties, I'm assigned the platoon sergeant position in the 406 General Supply Company at Fort Bragg, North Carolina. When I arrive at the reception center, I'm told the company commander wants to interview me.

That's odd. The typical procedure at reception is a review — your 201 file and your ERB, or enlisted record. There shouldn't be an interview.

I report dressed in my Class A, or "dress," uniform, with my drill sergeant badge, my recruiter badge, and my rows of ribbons.

The company commander comes out and introduces himself and his first sergeant. Then he says, "We just want to talk to you to make sure you can handle troops."

"Excuse me, sir?"

"I want to make sure you can handle troops."

I'm so pissed off it isn't funny.

"I'm a former drill sergeant," I say, "I'm a former recruiter. I just came from Korea, where I was the first and only female sergeant first class assigned to matériel support command, and I'm a member of the Sergeant Morales Club, which is awarded to active military members who promote the highest ideals of integrity, professionalism, and leadership while serving in Europe — and you want to know if I can handle troops? What kind of question is that?"

The company commander hems and haws.

My work experience and my nearly five years as a drill sergeant have given me so much more confidence. "If I was a man," I say, "would you be asking me that question?"

"We don't mean to insult you."

"Well, I feel insulted." I'm a little smart-mouthed sometimes, and I add, "If this is how you treat your senior NCOs, you and I are going to have a hard time. I'm a sergeant first class of the United States Army, sir."

A similar incident occurs when I become a battalion sergeant major at Fort Campbell, Kentucky. This will be my first battalion.

I report early, just to say hi. I'm house hunting and have a few more days before I sign in. I meet my XO and then the battalion commander. He looks me over. Sighs and shakes his head.

"What did I do to get the girl?"

"I don't know, sir. I didn't pick you, either."

He stares at me as if to say, *Who the hell do you think you're talking to?*

What I'm realizing is that the higher my rank, the smaller my world. The fishbowl I was living in has become a shot glass.

AMERICAN HEROES

Everyone is watching me and I'm feeling the pressure. In each new position that I'm the first woman to hold, I can't help but feel I'm representing *all* women, and that if I screw up, every woman who comes after me will pay the price.

"You've been without a battalion sergeant major for several months," I say. "You've had a first sergeant acting as a battalion sergeant major. Why don't you give me a chance to see what I can do before you judge?"

What I really want to say is what I'm thinking. *You're not going to beat me down. I'm just as good as you, so we'll just see who wins in the end.*

In 1992, Somali dictator Mohamed Siad Barree is overthrown in a military coup. As two powerful warlords fight for control, the country plunges into chaos and famine. Millions are starving to death. The United Nations organizes a humanitarian effort. The US sends its military to help deliver supplies and protect the aid workers.

By 1993, the situation has become unstable. There's a major uptick in fighting.

I'm deployed to Somalia as part of a logistical task force. An engineer company is attached to us. We'll be responsible for purifying and delivering water, petroleum oils and lubricants, bulk fuel, and line haul—tractor trailer trucks used to transport containers, bulk cargo, and liquids from the port in Mogadishu to forward areas.

I'm stationed at Hunter Base. Being in the 101st and having the background from the male drill sergeant who taught me infantry tactics—all that knowledge comes into play.

The battalion commander and I immediately realize that none of our facilities, not even our headquarters, are hardened—not a single sandbag anywhere—causing concern as there are high-rise buildings directly across the street. Anytime, the bad guys can look right down at us and see that we're not protected and start firing and lobbing RBGs into the compound.

"Granted, I'm a logistical person," I tell my commander, "but this lack of protection makes no sense to me."

He agrees. There's no way we can build up the walls around the compound, so we start sandbagging everything. Then we repurpose shipping containers to create bunkers.

I've been here a couple of days. There are still more changes I want to make.

We install guard towers and then decide where to place the .50-cal and M60 machine guns and Mark 19 grenade launchers. But who's going to man them? Our folks are busy driving supply trucks. We develop guard forces and provide personnel for the guards in the overwatch positions.

The Bangladesh infantry company is also stationed on our base. We all need post security. We have them man the towers—but all the overwatch positions with heavy firepower are manned by US personnel.

The Army hasn't provided us with enough regular radios, so I get the battalion commander to buy CB radios to install in our trucks. On daily runs throughout Somalia, we need a means to communicate with each other in the convoy to ensure everyone comes back in one piece. We don't have any infantry soldiers to protect our drivers. It's all on us.

AMERICAN HEROES

The commander is the only one who has an armored Humvee. The rest of us make do with civilian-style Commercial Utility Cargo Vehicles, or CUCVs. To make them semiprotected, I place sandbags on the floor. If I'm driving in one and a grenade rolls under the truck and detonates, I want the shrapnel to go into the sandbag, not up my ass.

The outgoing unit keeps saying, "I don't think all this stuff is necessary." A lot of other people — including my own soldiers — agree.

My third day at Hunter Base, a land mine blows up on one of the nearby roads outside the compound. It shakes everyone up. Now my soldiers understand the need for the extra security precautions.

At the end of the compound, we build a test firepit where soldiers make sure their weapons are working properly before they leave in their convoys. Conditions are dusty and dirty, and everyone is writing home asking family and friends to send nylons to keep the dust out of their weapons.

An infantry unit arrives at Hunter Base. One morning I wake up and see the infantry troops inside the compound doing a platoon run in formation.

"Sergeant Major," I say, "do you see those buildings right next to us? The bad guys are always in there. We try to clear those buildings every day, but there are too many bad guys, and every now and then we take on small arms fire — and we've had RPGs launched at us from there."

I explain how one landed on a building where I was living. I rolled out of my sack — I don't remember how I got my boots and Kevlar on—and I ran up to the roof dressed in my PT sleeping gear to find out where the attack came from.

"A platoon of fifty people running around is a big target," I explain, "and the bad guys only need to get lucky once. I only allow my troops to do PT in groups of no more than five. I'm not going to tell you how to do your job, but I think from a force protection issue, we need to limit our exposure."

He just stares at me.

I'm not going to get into a pissing contest with him, I keep telling myself.

Fortunately, he ends up coming around to my way of thinking.

UNOSOM II—the United Nations Operations in Somalia II—is tasked with the restoration of peace, stability, law, and order. The people at the higher levels talk with one another, but that information doesn't trickle down to the lower levels. It's a classic case of the right hand doesn't know what the left hand is doing.

On October 3, 1993, neither I nor anyone on the base knows that a Black Hawk helicopter has been shot down in Mogadishu. Outside our walls, we can hear a firefight and we can see tracer rounds and helicopters.

Our foreign partners, the Italians, come by the base for fuel, and that's when we learn something is, in fact, seriously wrong. But they don't know specifics.

Our base can get information from the American Forces Network (AFN), a government-sponsored TV and radio service for military service members serving overseas. I go upstairs to the one little TV we share, and on the news, I see the remains of one of our Army Rangers being dragged naked through the streets of Mogadishu.

I go to my battalion commander, tell him what I've seen, and say, "We need to get to the airport."

It's the location of our mortuary affairs point. As we're getting the convoys together, a call comes in that our chaplain is needed. We bring him with us and drop him off at the EVAC hospital.

A couple of remains are already there and casualties are incoming. Medical troops are doing triage out in front of the hospital, deciding who goes in first and who gets what.

We process the remains of eighteen Rangers.

In 1997, I'm asked to interview for a position at Fort Leavenworth, Kansas, where I would be responsible for all noncommissioned, officer, and warrant officer education. Four men also interview for the position.

They're not going to hire me.

General Meigs calls and tells me I got the job.

"I want you to know I didn't hire you because you're a woman," he adds.

"Well, that's good, sir, because I don't think I would take the job if you did."

"You're a soldier's soldier—a muddy boots soldier. You've been in divisions, done a lot of things, and understand a lot. I'm hiring you to be my sergeant major because of what you bring to the table."

I'm working at Fort Leavenworth when, in 2004, I get a call from Lieutenant General David Barno, the overall commander of all US and coalition military forces for the war in Afghanistan. He asks me if I'm interested in being his sergeant major for Combined Forces Command Afghanistan.

"I'm always up for an adventure," I say.

Combined Forces Command Afghanistan headquarters is at Camp Eggers in Kabul. A US general is the commander, a UK general the deputy commander. We're responsible for all US and coalition forces that number anywhere between twenty-two thousand and twenty-six thousand.

A lot of people want to overcomplicate the training, but it's simple. The Afghan National Army (ANA) soldiers need to be able to shoot, move, and communicate. They also need ammo discipline. When they no longer can see the bad guy, they need to stop shooting instead of emptying their magazine.

Then there's the monumental task of government reform. While the US, Brits, and French are training the ANA, the Germans are helping develop a local police force, and the Italians are trying to help the Afghans with their judicial system.

We establish the Kabul Military Training Center to teach the Afghans how to train their own. Once we get a unit up and running, we connect them with our Embedded Training Teams, or ETTs.

When led, the Afghans are good fighters. Where we run into trouble is when we fail to consider their culture.

We give the Afghans CONEX containers full of boots, but instead of distributing them, the boots sit in the containers as the soldiers run around barefoot. We build a compound to house them, and we also build latrines. They break in a matter of days because, instead of sitting on them, they stand on them. We give them Bobcats to dig trenches, but they can work faster using a pick and shovel. After a couple of months, the vehicles we've given the soldiers are all wrecked because we haven't taught them how to drive and service the vehicles.

US soldiers tasked with changing Afghans' way of life are often shocked at how their own lives have changed.

A young and naive female lieutenant colonel comes up to me and says, "When are we going to get a beauty shop at Camp Eggers?"

Her question nearly throws me over the edge.

"Ma'am," I say, "have you ever been outside the walls of Camp Eggers?"

"No, but I've been downtown."

I take her downrange with me, to one of the provincial reconstruction teams in Ghazni. When we land, she steps off the helicopter and tells me she needs to use the bathroom. I point to the wooden port-a-pots.

She comes back and says, "Those don't have any toilet paper."

I reach into my rucksack and pull out a roll.

Her eyes grow wide. "You *carry* toilet paper with you?"

"You never know where you're going to be."

"What are those things?" she asks, pointing to a series of long tubes sticking out of the ground at 45-degree angles.

"Those are piss tubes," I say. "That's what the guys use."

"This place is so primitive."

"Look around you. You're in Afghanistan. Rely on your survival skills instead of your creature comforts. So think twice before you ask me again about whether the beauty shop is coming to Camp Eggers because you, too, can be living in a luxurious spot like this."

Women, I explain to her—to everyone—can be as adaptable as men.

Still, seeing women in leadership positions is not the norm.

I go to see some Special Forces guys down in southern Afghanistan. We take a helicopter and then drive the rest of the way in a Humvee. I'm dressed in my "battle rattle"—my full combat gear.

When we reach the secure compound, I start removing my gear. There, the captain looks at me and says, "Oh, my God, you're a woman."

"Oh, my God," I reply, "how did that happen? *When* did it happen? Must've happened on the flight down."

The poor captain makes an *oh, crap* face.

I make a joke out of it. "Look, sir, I get it. You expected a crusty infantry, armor, or artillery sergeant major, not some female logistician. I get it. Let's move on."

We go through the briefings, and I ask a lot of questions and answer every question I'm asked.

The captain says, "You really do understand this."

"Yes, sir. I think that's why the general hired me."

Over the years, when I'm asked about integrating women into the Army and whether women should be in combat arms, I always give the same answer: whoever can meet the standard should be able to do the job. Don't change the standard. Just put the standard out there and let water seek its own level.

Women *are* in combat. What women are not in are in combat arms. They're not in infantry, armor, or artillery. That's a debate for another day. But the fact is, when women go defend the perimeter, they pick up their weapons and do what they have to do. When women operating a truck convoy are challenged by the enemy, they do what they have to do. When a rocket attack happens, the rocket is indiscriminate. It doesn't say, *Oh, look, there's a group of women, let me go over there.* It will hit where it's going to hit.

All women soldiers want is the same training given to our male counterparts. When women deploy, they need to know how to take care of themselves and their fellow comrades, be they male or female.

Women have perseverance. There's always going to be a ceiling somewhere we'll have to break and it's how we go about breaking it. Women in the military have gone about breaking these ceilings by showing what we can do, not asking for special privilege. Give us an obstacle, okay, we'll overcome it. When we overcome that obstacle, there'll another one. At some point you're going to run out of obstacles for us to overcome.

Ralph Puckett, Jr.

First Lieutenant, US Army
Conflict/Era: Korean War
Action Date: November 25–26, 1950
Medal of Honor

World War II is still raging. I'm sixteen, and I love combat movies—especially air combat. I watch them and keep asking myself, *Do I have what it takes to become a soldier and a military pilot? Can I measure up?*

I've done a lot of hard, physical work. I'm not afraid of it, and—so far—I've been able to withstand the physical and mental demands of some really tough jobs.

I want to be a pilot. I pay for my own flying lessons and get a pilot's license a year before I can legally drive.

It's 1943. I enlist as a private in the US Army Air Corps Reserve.

In 1945, I'm accepted to attend the United States Military Academy, West Point, where I eventually become captain of the boxing team. My desire as a cadet is to become an Army pilot. After my junior year (or "cow" year as we call it), I visit Fort Benning. I'm impressed by the infantry school's discipline. These guys train to standards and adhere to standards. That speaks to me.

I switch to infantry.

Colonel Lewis G. Mendez, a highly decorated World War II combat veteran, is an infantry officer assigned to West Point. As part of the Army's elite 82nd Airborne "All American Division," Mendez parachuted behind enemy lines on a daring mission critical to the D-Day assaults on the Normandy beaches, earning the Distinguished Service Cross and three Bronze stars.

I'm a great admirer, find the man inspiring. I'd be honored to know him. My roommate introduces us, and Mendez becomes my mentor.

It's 1949—my graduation year—and when I leave West Point, I know I'm eventually going to be assigned to Okinawa for occupational duty. When summer arrives and I'm going through infantry training, there are rumbles of war in Korea.

I volunteer to go. I want to lead troops.

As I go through infantry training, fear is on my mind. I have a lot of fear. Learning how to successfully manage it, I believe, is one of the most important aspects of being a soldier. I seek Colonel Mendez's guidance.

"Sir, I know I'm going to be afraid. How do I manage it?"

"Ralph, if you're afraid and thinking about yourself, you're not doing your job. Just get up and do something."

"Like what?"

"Check in with your men. Think about them, not yourself. That's your job."

I'm sent to a replacement depot in Japan. The war in Korea has started. On my last day before shipping out, I hear my name called over the intercom system.

I'm ordered to report to headquarters. I arrive and meet Lieutenant Colonel John H. McGee, a veteran of the US-led Philippine guerrilla campaign during World War II.

"I'm interviewing officers for assignment in a Ranger unit I'm forming," he says.

The Rangers are the Army's elite fighting force. They were deactivated after World War II. Korea created a new need for them.

"My problem, Lieutenant, is that I've already selected the two platoon leaders. There's only one officer slot left. Company commander."

"I volunteer," I tell the colonel. "I don't have any experience. I'll take any position you give me."

He interviews me at length, trying to figure out what I've accomplished, which isn't much. I've been off at school. I've completed three basic officer's courses. I'm airborne qualified. I'm a second lieutenant. That's it.

The colonel, though, seems impressed that I was on the boxing team at West Point. That I'm in outstanding physical shape. That I've gone through — and finished — jump school.

"Let me think about this," he tells me. "I'll call you Monday morning and tell you yes or no."

He calls me Monday, as promised. "Not only have I selected

you to be assigned to the Ranger company," he says. "I'm going to appoint you the Eighth Army Ranger Company Commander."

I feel a punch to the gut.

I don't have one ounce of experience. I also know I'm getting the best opportunity I will ever get in my life.

Dear God, don't let me get a bunch of good guys killed.

I arrive in Korea on September 2, 1950.

I know I am in way over my head. I know it's a great risk for the Army to have placed their faith in me. I know it's a great risk for my soldiers who are going to be Rangers.

I begin to recruit my Rangers but there's one small problem. I can't have any infantry soldiers in my unit because all infantry units are in short supply. I'll have to select soldiers from support troops — truck drivers, clerks, food service personnel, all non-infantry men, all inexperienced — and turn them into Rangers.

I personally interview each man and ask two questions. Are you willing to train to standard, and will you follow me? I select men based on their duty performance, athletic ability, weapons qualification scores — and their desire to serve as a Ranger.

I select seventy-five men. My team includes two Blacks, a Latino, and Asian American members. The Army has only recently been desegregated, but I never cared about color. We all bleed red.

We travel to a rice paddy in South Korea for seven weeks of infantry training — specifically Ranger training. I create what will eventually become the Army Ranger School curriculum: focusing on fundamentals, enforcing training standards, rotating leadership positions, after action lessons learned. Though strict and demanding, I also make sure to praise them for the good, hard work they're doing.

It's early November, the days cold and raw. Leadership says American forces will be home for Christmas, but I have my doubts. Throughout Korea, tough fighting — yielding tough casualties — continues.

Five-and-a-half weeks in, our training is cut short. We're going to war.

The Eighth Army Ranger Company is attached to Task Force Dolvon, the battalion-size armor lead element of the 25th Infantry Division. We begin to advance north. My Ranger unit leads the task force's assault. We ride, exposed, on the cold armored decks of Sherman tanks.

Conditions are brutal, and we're not properly dressed for the weather, wearing standard field jackets and regular boots. Strained supply lines struggle to get us proper winter gear and food and ammo. The ground is frozen, making it difficult for us to dig the foxholes where we sleep at night. Not that we get much sleep. Hypothermia stalks us constantly. To prevent frostbite, men remove their boots and stick their feet into a fellow soldier's armpits.

But we fight through it all and, on November 25 — Thanksgiving weekend — my Rangers are ordered to capture Hill 222. We successfully accomplish the mission, but not without taking casualties from both enemy and friendly fire. To get our tanks to cease firing on my Rangers, I'm forced to run across open terrain, jump on the top of the nearest Sherman, and bang on the turret hatch with the butt of my rifle to get their attention.

AMERICAN HEROES

The following morning, we receive intelligence reports that more than twenty-five thousand Chinese fighters are preparing to enter the war within our area of operation.

We receive new orders. Seize and defend Hill 205, a strategic position that overlooks the Chongchon River in North Korea. Sixty miles north is the Yalu River, the border between North Korea and China.

To accomplish this mission, I have only fifty-one US Army Rangers and nine Korean Augmentation to the United States Army (KATUSA) soldiers.

God, please let me take care of these men and don't get them killed.

Temperatures are well below freezing, and we're already exhausted and battle weary. Yet, in the early morning hours, we once again head out to fight, riding on top of the tanks.

As we approach our objective, we start taking mortar fire. The tanks stop and button up. I turn my attention to the eight hundred yards of wide-open terrain between our current position and Hill 205. We jump off and form up.

I lead the Rangers across frozen rice paddies. We split up to flank the enemy. One platoon gets pinned down from heavy gunfire coming from a hidden location. A machine gunner's nest is somewhere out there, and we need to locate it.

I decide to take a risk. I expose myself by running across the field. The gun fires but we can't spot its location. I find cover then make another run. The gun fires again. Still no location. On my third attempt, we finally locate it and eliminate the threat.

The tanks have yet to fire. I run across the open field, bang on a hatch again to get them firing, then lead my Rangers to the base of the hill and then up its rugged mountain face rising 205 meters high. Meeting minimal resistance, we secure the hilltop. Hill 205 is ours.

Our tanks are at the hill's bottom, unable to support us on the top. The closest friendly unit is well over a mile away. We're armed with our individual weapons, machine guns, rocket launchers, and grenades. In taking the hill, I lose six Rangers and three KATUSA—Korean enlisted personnel assigned to my unit—which leaves me with fifty-one total. That's it, that's all we've got.

I feel all alone, but totally focused on my direct responsibilities. Now we need to hold the high ground. Night is falling and we begin work to establish a strong line of defense against an enemy counterattack we know is coming.

Confident that my defensive plans will be implemented, I head down the hill and back to battalion headquarters with a handful of my Rangers to secure supplies and coordinate artillery fire support.

It's around ten o'clock and pitch black when I make it back to my company lines. The temperature is in the upper teens, but with the wind chill factor it's below zero. I've just entered the perimeter when I hear bugles and whistles in the distance.

Then, from the cold darkness, we hear four chilling words repeatedly blared over the enemy's portable speakers.

"Tonight, you die, Yankees!"

AMERICAN HEROES

We'd fought the North Koreans to take the hill. Now we're going to face the Communist Chinese who want it back.

Mortar rounds begin to explode all around us as the enemy lays down heavy machine-gun fire. I take shrapnel to my thigh and crawl back into my foxhole. One of my platoon leaders is there. He wants to evacuate me.

I refuse. Fifty-one men are here under my command — fifty-one good men. They're an inspiration. They weren't originally infantry, have had only a few weeks of learning basic combat skills, and yet they're tough and motivated to fight. I need to look after them, help them hold the line. They're my Rangers, my command.

We're facing an enemy battalion of six hundred infantry, so we're outnumbered twelve to one. I start to call in artillery fires in accordance with the indirect fire plan I'd coordinated earlier at battalion headquarters. Artillery is our only chance to break up the overwhelming human wave attacks we're about to face. Artillery is our only chance to survive.

My God, I'm afraid.

I recall Colonel Mendez's advice back at West Point.

Stop thinking about yourself. Get up and do something. Go check on the men. What can you do to help them do their job?

When I'm not calling in "danger close" fire missions I'm running from foxhole to foxhole, exposing myself to enemy fires to deliver supplies, check the company's perimeter, and look in on each man.

One of my Rangers points in the direction of an enemy sniper who's taking shots at us under the cover of darkness.

Where is he? I need to draw him out. I jump out of a foxhole, making myself a target, hoping the sniper will fire before I throw myself to the ground.

After three such attempts and exchanges of gunfire, the sniper threat is terminated.

The enemy's massed waves of humanity keep attacking. Our numbers are dwindling. We're almost out of ammo. I'm wounded again, this time in the shoulder. I push past the pain. I refuse to be evacuated. I'm not leaving my men. I distribute my ammunition, keeping only one eight-round clip for myself.

Finally, after five unsuccessful attacks, the Chinese change their tactics from broad assaults to a pinpoint target. Massing in the darkness, they start placing heavy fires on a distant corner of the perimeter. They advance, overwhelming the few Ranger defenders located there.

As the enemy penetrates our defenses, I move as quickly as possible back to my foxhole and request artillery support.

"Sir, I can't help you now. The entire front is under attack. I'm involved in a different fire mission."

"We've got to have it," I say. "We are under heavy pressure."

I order my remaining Rangers to fix bayonets. It's 2:35 a.m. It's the enemy's sixth assault. I know this is going to be our last stand.

Heavy enemy mortar fire continues to rain down on our positions as they press through the perimeter hole. Two mortar rounds explode in my foxhole. I'm brutally slammed against the dirt wall, my right foot nearly blown off, my entire backside peppered with shrapnel. I can't move. I can't do anything.

The platoon leader with me—a West Point classmate—took the brunt of the blast. There's nothing left of him. My radio has survived, and I make one last desperate call.

Still no available artillery. I tell the battalion we're done, being overrun.

Somehow, I manage to crawl out of the foxhole and look around. My Rangers are being overwhelmed by the enemy who are everywhere. Some are engaged in hand-to-hand combat. I see three Chinese about fifteen yards away from me bayoneting or shooting some of my wounded Rangers.

Be there. That has always been my personal standard. Be there for my soldiers, and if a soldier is undergoing adversity, be there by his side. But I can't move, and I can't help them. I order everyone to evacuate.

A Ranger approaches out of the darkness. He's unarmed. Unable to move, I order him to leave me behind. He leaves but informs two armed Rangers below who charge up the hill. They shoot and kill the three Chinese soldiers nearby.

"Sir," one of the Rangers says, "are you hurt?"

I'm hurt bad and the Chinese are swarming across the hill. I order the two Rangers to save themselves and leave me behind.

They disobey my order. "Sir, we're taking you with us."

One Ranger puts me over his shoulders, the other provides covering fire. As we stagger down the hill, bullets zip by us out of the dark. Exhausted and unable to carry me any farther, the Ranger places me on the ground. Again, I tell them to leave me, save themselves but, again, they refuse my order, dragging me down the remainder of the hill on my back to the Task Force Dolvon tanks below.

The pain is overwhelming. Dizzying. "I'm a Ranger," I mutter out loud, to myself and the others. I can take the pain. "I'm a Ranger."

That morning, I started out with fifty-one Rangers and soldiers. Now I have twenty-two standing in ranks, fit for duty. The rest are dead, wounded, or missing.

Before I'm loaded on the back of a tank along with other wounded Rangers, I direct the tank commander to call in a final concentration of artillery fire on Hill 205 as a parting gift.

I'm evacuated to the Fort Benning Hospital, which is near Tifton, Georgia, my home. Rehabilitation from my severe wounds—especially my nearly amputated foot—is going to take months, possibly even a year.

"We're going to get you patched up as soon as we can," the doctor tells me. "Then we'll get you medically discharged so you can go home."

"I'm not taking a discharge. I'm staying in the Army. I'm infantry."

My name and picture appear in the local newspaper. One afternoon, two young women appear in my hospital room, a college freshman and high school senior, whose typing teacher, a seventh-grade teacher of mine, had shown her my picture from the paper and asked her to stop by.

The high school senior's name is Jean. We have a nice conversation. Jeannie begins to visit more often—two afternoons a week become weekends—a steady thing. We fall in love. We want to get married.

Her father is against it. He wants Jean to go off to college. She does, but after her first year, we still want to get married.

I need to be honest with her.

"I will take you away from your family," I tell her. "You'll be living by yourself. You'll be raising children by yourself. I may be killed in action. You'll have to take over everything."

"That doesn't matter," she replies. "I can do it."

Two years later, we get married in Georgia, on the anniversary of "The Battle of Hill 205." She's a true Ranger wife.

I'm awarded the Distinguished Service Cross for my actions in Korea. In 1967, I'm awarded a second, along with two Silver Stars, as a battalion commander in the Vietnam Conflict. Those awards, in addition to five Purple Hearts, and two Bronze Stars—and my later inauguration into the Ranger Hall of Fame—all the credit goes to my sergeants and my men. I know it was tough for them, but I was right there with them, and I tried to praise each man every chance I got.

That's my leadership style. Just be there. No great plans of maneuver or anything like that. I've never been anything much more than another rifleman.

In the early 1990s, I get a call from Lieutenant Colonel (Ret.) John Lock, a Ranger who is also a military historian. He asks for information on the Eighth Army Ranger Company and the battle of Hill 205 for a book he was writing on Ranger history.

I answer his questions. Lock says he feels my actions meet

the requirements for a Medal of Honor. I'm not looking for any recognition, but Lock seems hell-bent on making his case.

Lock starts his quest in 2003, submitting the upgrade packet in 2004. The upgrade is denied in 2007, as is an appeal in 2009.

I beg Lock to stop wasting his time, but he presses on with subsequent appeals and pressure, finally achieving success in April 2021 when I receive a call from President Biden that my Korean Conflict Distinguished Service Cross had been upgraded to the Medal of Honor.

I'm told I'll be going to the White House to receive the award. South Korea's president Moon Jae-in will be attending and giving a speech. He'll be the first foreign leader to participate in a Medal of Honor ceremony.

"Why all the fuss?" I ask. "Can't they just mail it to me?"

On May 21, 2021, I attend the ceremony with my wife, two children, and six grandchildren. A granddaughter is an Army captain and an artillery officer. I'm ninety-four then, and seventy-plus years ago when I joined, women weren't even allowed in the regular Army.

Now women serve in combat branches. In 2015, the first women start to attend the elite US Army Ranger School. I'm all for it—and have said so to anyone who asks. We need them. They'll do a good job. And most will be outstanding because they've measured up to the Army's highest standards.

As Presidents Biden and Moon talk about the battle on Hill 205, my service in Vietnam and, after my retirement from active duty, serving as the Honorary Colonel of the 75th Ranger Regiment, mentoring new generations of Rangers and soldiers, my mind keeps going back to those seventy-five Rangers I selected, trained, and served with in Korea.

I'm proud, very proud of these Rangers. They were trained to be physically, mentally, and morally tough. They were highly skilled as a small combat unit and they made me believe that ours was the best company in the Army, a fact clearly demonstrated by them against overwhelming odds, under the most adverse conditions, deep within enemy territory that cold night on Hill 205.

My Rangers deserve this award. They did the training. They did the fighting. They did the dying. Two of them carried me off the battlefield. Those men are ones who should get the credit. They're the ones who earned this Medal of Honor.

RLTW. Rangers Lead the Way.

Duane Edgar Dewey

Corporal, US Marine Corps Reserve
Conflict/Era: Korean War
Action Date: April 16, 1952
Medal of Honor

Interview with Arline Broome, daughter

"Here's the medal my dad got when he was in the Marines," I say when, as a kid, I open my dad's dresser drawer and pull out his fancy medal to show to my friends. "What did your dad get?"

The only time I see my dad, Duane Edgar Dewey, wear this medal is when he puts on his old Marine Corps uniform to lead the annual Memorial Day Parade in South Haven, Michigan, where we live. He never talks about his medal; never tells me or my younger brother Dwight why he has it.

What my dad does tell us about is how he grew up poor on a farm in rural Kalkaska, in a one-hundred-year-old house without any running water or electricity. In the winter, Dad said, he and his sister would bring their shoes to bed at night, to prevent them filling with the snow that came in through gaps in the siding between the boards of the house. When he's twelve years old, the family moves to a housing project in Muskegon, Michigan. It's a one-bedroom apartment but seems very luxurious in comparison, with all the modern conveniences that were lacking in their previous home.

At age sixteen, my dad quits school and moves to South Haven, on Lake Michigan, to live with his aunt and uncle's family and work on fruit farms. When the Korean War breaks out in 1950, Dad notices that many of his friends and family are either getting drafted or enlisting.

One day, a cousin comes into Dad's room and says, "Let's take a little drive over to Kalamazoo."

"For what?"

"We're going to look into the Marines."

"I'll go, but I'm not joining," my dad tells his cousin. But about three hours later, he changes his mind and decides to sign up.

Dad's nineteen when he joins the Marines, and marries my mom, Bertha. After training at Parris Island and Camp Pendleton, he ships out to Korea in the fall of 1951, shortly before I'm born.

Dad—Corporal Dewey—is the leader of a machine-gun squad in Easy Company, part of the 1st Marine Division, 2nd Battalion, 5th Marines, which in the spring of 1952 is stationed near Panmunjom, on the border between North and South Korea. The command establishes a series of outposts beyond the main American forces, and Dad's platoon is dug in at one of these positions in the early hours of April 16, 1952, when they're attacked by a battalion-size Chinese force of between five hundred and seven hundred soldiers.

It would be several decades—almost half a century—before I'd hear my dad tell the story of what happens next in his own words:

My first ammo carrier and I go on watch. I duck down in my foxhole to light a cigarette.

Something goes over our heads. BOOM.

"I think that was a grenade," I tell him. Something else flies over our heads. BANG. I say, "Now I know it's grenades."

We start lobbing grenades back. Then all hell breaks loose. I take the cover off the machine gun. My gunner and assistant gunner are sleeping. I send my first ammo carrier to wake them up.

I start spraying the enemy. We're surrounded by nearly seven hundred Chinese soldiers. There's no place to go.

The firefight goes full blast for a little while and then there's a little quiet time and then BANG again.

My ammo carriers have M2 carbines, and we run out.

"I'm going back to see if I can get some ammo," I say, taking my gunner's .45 and heading off.

When I reach the ammo dump, I'm given a box of loose ammo for the carbines.

"I could use a case of grenades," I say. He gives me two hand grenades.

"Make them count," he says. "That's all we have."

I run back to my foxhole, hollering, "This is Dewey. Don't shoot me." *Our attackers are so close, I don't want them to overhear our real password.*

"There's a burp gunner down there in the barbed wire," my gunner tells me. "And every time I open up, he sprays us. Why don't you see if you can pick him off with your M-1."

"I just lobbed a grenade down there," I respond.

It goes off. "You got it right down his muzzle," my gunner says.

My gunny sergeant tells us to pull back to a smaller perimeter. There've been a lot of casualties. I don't like that idea very well, since now we've got no foxholes, but I find a big rock and tell the gunner to set his weapon up alongside it to give us a little bit of protection. We're down to three cans of ammo. Not very much for a machine gun.

Once again, I take my gunner's .45 to see if I can scrounge up some more ammo. No luck. I get back to my position and I'm still upright when a grenade goes off behind my left heel. I get shrapnel in my leg and left buttocks. It puts me down.

"Take over the squad," I holler to the gunner. "I've been hit."

A hospital corpsman arrives and asks where I've been hit. I tell him, and as he's getting my britches undone to examine the injury, a second grenade rolls beside me.

I grab it and I'm going to throw it. First impulse to get rid of it, right? But I'm lying flat on my back and I'm thinking I can't get this out of reach of my own men. So I scoop the grenade under my right hip and grab the corpsman.

As I pull him down on top of me, I say, "Hit the dirt, doc. I've got it in my hip pocket."

The grenade goes off. Takes us both off the ground. My body absorbs the full force of the explosion.

My next words are "Get me the hell out of here, I can't take much more of this."

The corpsman and the gunny sergeant drag me to a bunker full of wounded men.

As I'm given a shot of morphine, I think, "Okay, this is it. We're going to bleed to death, or the Chinese troops are going to come in here and finish us off. I'm looking Old Man Death right in the face."

I don't pray for myself. I spend the rest of the night praying for my wife, Bertha, and my infant daughter, Arline. She was born right after I left for Korea. I pray that Bertha will find a good father for our daughter and a good husband for herself.

Near daybreak, someone pokes their head inside and tells us the Chinese have pulled out. I'm sent to a field hospital, where I'm told I've also taken a bullet to the stomach. The blast put a good-size hole in my hip, but it missed my spine. I know how lucky I am to be alive.

My dad's act of heroism is dated April 16, 1952. Of the eighty Marines of Easy Company who defended that outpost in Panmunjom, thirty-six are killed or wounded, and two captured. Easy Company soldiers also earn one Medal of Honor, one Silver Star, three Navy Crosses, and thirty-six Purple Hearts that night.

My dad is awarded the Purple Heart for being wounded in action, and after healing for approximately four months in various military hospitals, he is discharged from active duty, with no military benefits. He returns home to my mom and me in South Haven and learns to cope with his injuries — including recurrent back problems and a hip that's always going out of joint.

Dad joins several family members working on the production line at Everett Piano Factory, even though the work's difficult for him. During one shift, he gets summoned to the office for a long-distance phone call from Washington, DC.

"It was someone from the White House," he tells his coworkers when he returns. "I'm going to be awarded the Medal of Honor."

These guys, many of them also veterans of either the Korean War or World War II, are deeply impressed. They congratulate him on the huge honor, carrying on to the point that their talk brings the supervisor to the floor.

"Duane," he tells my dad, "you're going to have to go home."

"What? Why?"

"You're disrupting the line and we're not getting any work done. Just go home."

My dad's coworkers try to figure out something special to do for him and come up with an idea. They set up a fundraiser, and soon everyone in town is involved. Through the "Dewey Fund" enough money is raised to purchase a piece of property and a pre-manufactured house. Many volunteers manage to erect the three-bedroom house and completely furnish it during the six days that my parents and grandparents are away in Washington, DC, for the Medal of Honor award ceremony.

Dad receives the Medal of Honor at the White House on March 12, 1953. He's the first soldier to receive one from President Dwight D. Eisenhower, who looks at him and says, "You must have a body of steel!" after reading the citation.

Upon the family's return to South Haven, there's a major

celebration, complete with many distinguished guests, a Marine and police escort, a reception and luncheon program, a parade, a City Citation with a Key to the City, and the presentation of the keys and deed to our new home. All witnessed by numerous media outlets.

The next week, my dad returns to his job on the line at the piano factory. His Medal of Honor is put in his sock drawer, and only comes out when he wears it for the Memorial Day parade.

In 1959, Dad decides to take advantage of vocational training for veterans with the goal of becoming a barber. Because of his hip injuries he is discouraged from barbering (too much standing) and encouraged to find something where he can sit down to work. He's pushed to learn office machine repairs. It clicks. After eighteen months of training, he decides to open his own business, South Haven Office Machines. The family garage becomes his office, the home phone is also his business phone. Since he can set his own schedule, he also drives a school bus in the morning and afternoon.

While I'm growing up, my dad never talks about his Medal of Honor. I don't find any books back then to explain what the medal means. I knew my parents were invited to the White House in 1963, but I hear more about it from my friends' parents than I ever do from my own.

It's not until 1968, in my sophomore-year high school history class, that I finally learn more about the military and what military awards mean. The Vietnam War is a constant topic of conversation at this time, with a lot of my high school classmates either leaving school to join the military or trying to figure out how to possibly avoid the upcoming draft.

My history teacher knows my father personally, so takes the time to explain the significance of the Medal of Honor to our class while discussing the history of the Korean conflict.

"Arline, could you get your dad to come speak at school?" my teacher asks.

"I'm not a public speaker," Dad says when I ask him, declining even when I offer to write a speech, or do the talking. "I won't be part of show and tell, either," he tells me.

So, no personal words from my dad to my high school classmates.

Dad is a very shy person, but he enjoys spending time with fellow veterans at the VFW and the American Legion. He's dedicated to helping veterans, and becomes the service officer for the VFW, including a term as commander of the local post. It's at the VFW that he first learns of Glenn Higgs, a local boy who'd left school to join the Marines, like Dad, and was then wounded in Vietnam. Glenn stepped on a land mine—and as he was being carried out on a stretcher, a mortar blast inflicted even more wounds, filling his body with shrapnel. On the same day his fellow classmates graduated high school, Glenn lay in a hospital bed in Japan, dealing with the loss of his legs.

My dad hears that since returning home, Glenn hasn't been doing too well, that he's bitter and depressed. Dad visits Glenn at home. What begins as practical concern—explaining to Glenn that he's entitled to benefits, then helping him complete the paperwork to get them—grows into a close friendship between the two of them, filled with many years of hunting and fishing trips.

By this time, I've moved to Florida, where in 1972, I am blessed with the birth of twin sons, Lenny and Denny. Not long after, my parents sell their house in South Haven and move two and a half hours north to live in their rustic hunting cabin in Irons, Michigan. They also make their first trip down to Florida to see their grandsons. Dad falls in love with the weather, so they begin spending every winter in Florida, and as much time as possible with their grandsons.

My mom collects newspaper articles over the years about my dad's military service and honors and spends hours and hours creating scrapbooks. My sons grow up with a much better appreciation of their grandfather's military service and the significance of the Medal of Honor than myself or my brother ever had when we were younger.

As a Medal of Honor recipient, my father is invited to all presidential inaugurations. In 1989, when President George H. W. Bush is inaugurated, my mother steps aside to allow me to accompany Dad in her place.

I'm thrilled to have my first opportunity to meet other Medal of Honor recipients like my dad. I bring with me a book of Medal of Honor citations printed by the Library of Congress and including citations for each medal awarded going back to the Civil War. Though it contains no personal information, and no photos, the book gives me a great opening to meet many of the other Medal of Honor recipients in attendance and ask them for their autographs.

At one point, I get into an elevator with my dad and a man who I see is also wearing a Medal of Honor.

"Excuse me, sir," I say, holding up my book and introducing myself. "Can you tell me which one of the recipients you are?"

"I'm Drew Dix."

I flip to his page and read his citation, awarded for bravery during the Vietnam War. But there's a star next to his name, indicating he is deceased. I look up at him.

"This book says you're deceased."

"I know," he says. "You can't rely on the government to get everything right."

We all chuckle as he signs my book.

Later that same year, my sons Lenny and Denny, then in high school, discuss different branches of the service with their granddad and what each could offer. Denny signs up for deferred enlistment in the Navy. He feels like that's close enough to the Marines. Lenny also decides to do deferred enlistment, but initially chooses the Air Force.

"That's a good move," Dad says. "You'll learn a lot of things that you can use later on in life."

But Lenny changes his mind about the Air Force and ends up committing to the Marines. Right after the boys graduate high school in 1990, they both enter boot camp.

At boot camp, Lenny hears a lot of talk about Medal of Honor recipients—the highest honor anyone in the service can earn. Lenny tells his commanding officer that his grandfather is a Medal of Honor recipient. The commanding officer doesn't believe him until Lenny shows his CO one of the scrapbooks he asks me to send him from home. His commanding officer then asks Lenny to invite his grandfather to boot camp graduation.

Dad, who is extremely proud of both his grandsons, agrees. When graduation day arrives, a big deal is made of Dad being there. Between that and the heavy rain that forces the ceremony into much closer quarters inside instead of out on the parade field, I feel a little sorry for Lenny's graduating unit, but I think everyone there thoroughly enjoyed the day.

In 2000, I move back to Michigan and start a job with the City of Holland, though all my family and friends are in my old hometown of South Haven. Much to my surprise and delight, I discover that the house where I grew up is on the market. I buy it, and still live here today.

Dad rarely feels comfortable talking about his time in Korea or his Medal of Honor until 2010, when he's invited to Gainesville, Texas—aka "The Medal of Honor Host City." Medal of Honor recipients are invited there to attend various celebrations, where they are introduced to and speak with city residents and are also taken to several local schools to meet students. The students go above and beyond in welcoming the recipients with personal escorts, patriotic decorations throughout the schools, and one-on-one interactions. The recipients sit in front of the auditorium or gymnasium and answer student questions or go in groups of two or three to visit different classrooms and briefly share their stories.

Forty years after I asked him to come speak to my history class, Dad's finally in a place where he's willing to talk about his time in Korea and the Medal of Honor. From then until his health starts to fail, he returns to the event in Texas every year and also begins attending annual "Lest We Forget" gatherings in Benton Harbor, Michigan, to share his story with groups of adults. He also agrees to engage in multiple interviews and recordings about his experiences, as well as making several appearances with different military associations.

My sons and I are often asked what it was like growing up with my dad, their granddad. To me, he is just a normal dad. To my boys, he is just their granddad. We are very proud of him, and it is wonderful to attend celebrations where he is recognized. We try to answer any questions asked if they arise. But, just like Dad, we're not the ones starting any conversations about his medal or what he has done. In his own words: "The Medal of Honor means a lot to me, but every time I put it on, I think about other Marines who deserve the award and didn't get it. I didn't do anything that somebody else in my position wouldn't have done."

We are on a first-name basis with many of the Medal of Honor recipients. When you're with them, they're just ordinary people. It's easy to forget how special they are to the rest of the world, but I try my best to never let that happen. I know they are all special. I know they each did something phenomenal.

Alwyn Cashe

Sergeant First Class, US Army
Conflict/Era: War on Terrorism (Iraq)
Action Date: October 17, 2005
Medal of Honor

Interview with Tamara Cashe, widow

Even now, after all these years, it's hard for me to discuss my husband, Alwyn Cashe. I'm intensely private. I basically keep to myself.

When people ask me about Alwyn, what he was like, my fondest memory of his life, our life together, I don't have a particular story. Alwyn and I were two regular boring people living a regular boring life.

I've never been good with details—but I'm better with places than dates.

Our story starts at Fort Lewis in Seattle, Washington. We're both in the Army. I'm a cook. Alwyn joined in 1988, right after high school. He comes from a blended family, the youngest of nine kids. His family grew up poor, in a three-bedroom apartment owned by the housing authority, in Oviedo, a suburb of Orlando, Florida. He was five when his father died while undergoing surgery.

Alwyn and I fall in love and get married. I give birth to a daughter. We have no extended family nearby.

We're both serving in the same unit when, on August 2, 1990, Iraq invades Kuwait. This action leads to what will become known the following year as the Gulf War.

It's likely that my husband and I will both get called overseas. If that happens, we'll have no choice but to leave our newborn daughter with a stranger until a family member can come get her. I'm not comfortable with that, so I decide I'm going to leave the Army to raise our daughter.

In 1991, Alwyn deploys to Iraq as part of Operation Desert Storm. I'm now a single parent, a circumstance that is a part of military life. I don't expect — or want — people to feel bad for me. Besides, there are other jobs where families make similar sacrifices.

The Army is a melting pot of people from all walks of life, cultures, and social norms. Family Readiness Groups, or FRGs, is an Army program where soldiers, volunteers, civilian and family members — immediate and extended — form a network where everyone comes together to help each other out.

There's truth to the cliché about the military being a brotherhood. It's a great experience for our daughter. She learns to socialize with all types of people. As she gets older, she'll see that we're all the same.

It's a family business, being in the Army. We're all in this together.

My family moves every two years, to all sorts of places. When Alwyn returns from Iraq, we go to Korea and then Germany.

We're living in England when two planes fly into the World Trade Center. Everyone knows it's only a matter of time before we go to war.

Alwyn is sent back to Iraq in 2003. Before he leaves, his older sister Kasinal says, "Don't go over there playing a hero. You learn how to duck and come home."

The siblings disagree.

"I'm doing the job I was trained to do," he says. "I have to take care of my boys."

Alwyn cares for his men even when they're not on the battlefield. When soldiers need to take tests to move up in rank, Alwyn and I always make study cards for them.

While my husband is overseas, I don't worry too much about him. If anyone can figure a way out of a crazy situation, it's Alwyn.

He sends me a few letters. He's not much of a writer.

Alwyn and I don't talk much, either. Cell phones are new and not everyone has one, so the only way to call back home is to stand in line and wait your turn to make a call usually lasting no more than ten minutes.

He returns home for a few months and then, in 2005, is sent back to Iraq as part of Operation Iraqi Freedom. He'll be serving as a platoon sergeant in the 3rd Brigade, 3rd Infantry Division.

This time, he brings a cell phone. We speak once a week or sometimes once every two weeks.

Alwyn has about five years until he can retire — not that I think he will. Not unless the Army makes him.

I speak to Alwyn on October 17. I can't remember what we spoke about, but the conversation is uneventful. I have no idea he's about to go out on a mission. Or that this is the last time I'll speak to him.

Later that night, or maybe early the next day — I can't remember — I get a phone call. The man on the other end of the line says that my thirty-five-year-old husband is being flown to an Army base in Germany. Alwyn has burns covering nearly 72 percent of his body.

I enter a state of shock.

The man continues speaking, giving me details I can barely take in, let alone process. From Germany, he'll be flown to Brooke Army Medical Center in San Antonio, Texas.

"I know you have a daughter," he says, snapping me back to the present.

"We do." She's eleven. She'll be twelve in December.

"I wouldn't bring her," he says. "They won't allow a child of her age into the hospital room."

I fly to Texas, alone. There, I learn what happened to my husband.

From Forward Operating Base Mackenzie, a convoy of two Bradley Fighting Vehicles (BFV) containing seventeen soldiers and their interpreter sets off in the middle of a sandstorm.

On the night of October 17, 2005, Alwyn and his unit are leaving the run-down airfield north of the Tigris River to secure a supply route near Balad Air Base, a place company commander Colonel Jimmy Hathaway calls a "powder keg" where "there's always a fight going on someplace."

Alwyn's Bradley is in the lead. The tracked, medium-armored vehicle is equipped with a 25 mm gun and an antitank missile launcher, but because of the storm, they don't have any air support.

A couple of miles into their trip, the convoy takes on small-arms fire. Then Alwyn's Bradley hits a roadside bomb. The force of the explosion ruptures the vehicle's fuel tank. Jet fuel sprays through blast holes in the Bradley's hull.

The interpreter, seated in the back and covered in fuel, is immediately engulfed in flames. So is Sergeant Gary Mills, whose hands are too badly burned to open the rear door.

Staff Sergeant Douglas Dodge regains consciousness and finds he's trapped in the back of a burning vehicle. He reaches for the rear door handle. It burns his hand.

The interpreter is on fire and screaming. His friends are on fire and screaming. The Bradley is packed with combustible ammo. Once it ignites, every one of its trapped passengers will be dead.

Dodge scrambles for the breaching tool, uses it to force open the hatch. He's outside, on the ground and vomiting, when a voice yells, *"Dodge! Where are the boys?"*

It's Alwyn. He was seated in the gunner's hatch when they hit the IED. He managed to escape with minor injuries and then helped the driver to extinguish his burning uniform.

Alwyn's uniform is soaked with fuel—and it's beginning to melt. Without a moment's hesitation, he opens the rear door and moves through the flames, catching on fire. He grabs Mills, drags him out of the vehicle, then turns back to the flames and smoke to rescue another one of his boys.

"My little brother lived by the code that you never leave your soldiers behind," his sister will later say. "That wasn't just something from a movie. He lived it."

First Lieutenant Leon Matthias is seated inside the second Bradley, watching as Alwyn takes on small-arms fire. Matthias calls the base, requesting immediate aid, and has his men open fire on the enemy's trench line. Soldiers bolt to the wounded, to help smother the flames.

Alwyn rescues six of his boys. They're all badly burned. Alwyn is, too.

The casualties are delivered by convoy back to the base. Alwyn, suffering from second- and third-degree burns over three-fourths of his body and in unimaginable pain, insists his soldiers be evacuated first.

"It was the last time I saw him," Matthias later says. "Him walking to the helicopter in a shredded uniform."

No one knows how their mind and body will react until they're placed in a situation, but if I were on fire...I don't know what I would do. I don't know that I would be able to help anyone.

Now no one can help Alwyn.

It goes on for the longest time. He won't wake up. He must be sedated, but I don't know for sure because I'm still in a state of shock. People are dying all around me. Everything is happening so fast. I feel like I'm stuck on a big crazy hamster wheel.

A person gives me my husband's ring, saying, "You're going to have to wash it."

He's right. The ring melted into his skin.

Alwyn's wedding band is the only possession of his that made it home.

His older sister Kasinal is also at his bedside. She knew nothing about the explosion and Alwyn's bravery until a nurse says, "You know your brother's a hero, right?"

She knows. Before Alwyn left for his final mission, she told him not to play the hero, though she also knew he wouldn't listen. He would do anything to protect his boys.

Alwyn's boys are also here in Texas, being treated for their burns and wounds. Four don't make it.

One day, Alwyn wakes up. He sees his sister.

"How are my boys?" he asks.

She tells him.

Alwyn breaks down. "I couldn't get to them fast enough," he says.

He manages to get out of bed. Mills, who was placed in a coma for four days, has burns covering 17 percent of his body. But he's alive, and Alwyn is happy to see him. My husband is optimistic about his recovery. They make plans to go hunting together.

"I was standing there and talking to him," Mills says later, "and I was like, 'How are you not in this bad mood?'"

He's still in an immense amount of pain and the death of his boys weighs heavily on him.

Some of the guys from Fort Benning also visit that day. Alwyn manages to talk to them for a bit, and then he goes back to bed and falls asleep.

When I visit my husband the next day, I notice his toes are black. Alwyn, I'm told, is suffering from Mucor disease. It's a rare fungal infection caused by air- and earthborn molds called mucormycetes. The typically harmless spores can enter the body through a cut or an open wound, leading to serious infection in a person, like Alwyn, with a weakened immune system. They're going to have to amputate his legs.

Soon after the procedure, I realize my husband isn't going to recover from his injuries. The medical staff says it's time to let him go.

I send for my daughter. She arrives on November 8, 2005. She joins us at Alwyn's bedside and gets to see her father pass, almost three weeks after the attack in Iraq.

Three days later, on November 11, my husband is awarded the Silver Star, the Army's third-highest award for valor in combat. He's also awarded the Purple Heart, which is given to soldiers who are wounded or killed while serving.

Alwyn's father is buried in a little cemetery in Sanford, Florida. I bury my husband there so his mother can see her son whenever she wants. I want to make it easy for her.

In 2020, I receive a call from Vice President Mike Pence that my deceased husband has been nominated for the Medal of Honor.

It's taken sixteen years, but it's not for lack of trying. Alwyn's battalion commander, Gary Brito, started the recommendation process months after he learned the full details of Alwyn's selfless actions.

Unfortunately, Brito ran into roadblocks. Certain key details were lost during the chaos of that night given that most of the soldiers Alwyn rescued were in critical condition and couldn't provide firsthand accounts before they died.

Brito, though, didn't give up and embarked on a second application packet. While he worked on obtaining sworn statements from soldiers, Mills—one of the soldiers my husband rescued—and my sister-in-law embarked on a public campaign. They, along with Alwyn's fellow soldiers, commanders, and two high-ranking generals, broadcast my husband's valor.

Now Alwyn Cashe will be the first African American to be awarded the Medal of Honor in the wars following the terrorist attack on September 11, 2001.

"He is not just a black soldier who earned the right to the Medal of Honor," his sister says. "He's a soldier who happens to be black."

She joins me at the December 2021 ceremony, delayed because of the January 6 riots. Alwyn is one of three soldiers awarded the medal.

Because it's the military, everything is planned out—be here at this hour, there at that hour. Soldiers and their families typically have a meal with the president, but because of COVID that doesn't happen. Instead, we briefly meet President Biden and his wife right before the ceremony.

They're very nice—the ceremony is nice—and I'm thinking of Alwyn, wishing he were here.

If Alwyn had gotten out of the Army, he would have become a game warden, something like that. He loved hunting and fishing. He also loved helping animals as much as he loved helping people. One time I came home to find a deer in my bathtub. The year Fort Benning had an overabundance of wild pigs, I found one in our backyard standing in the back of his parked truck.

With Alwyn, you never knew.

Harvey Curtiss "Barney" Barnum, Jr.

First Lieutenant, US Marines Corps
Conflict/Era: Vietnam War
Action Date: December 18, 1965
Medal of Honor

Junior and senior boys pack the Cheshire High School auditorium for our annual military career day. A representative from each branch of the US Armed Forces is seated on the stage, each waiting for their turn to speak.

We're an unruly bunch. When the Air Force recruiter speaks about the virtues of serving our country, catcalls, hoots, and whistles erupt from the crowd. The same thing happens when the Army and Navy recruiters take the stage.

An old, crusty gunnery sergeant is the last to speak. He slaps his fist against the table and glares at us.

"There's no one in this room worthy of being a United States Marine," the gunnery sergeant says, his voice booming through the auditorium. *"I'm deplored that the faculty in the back of the room would let the students carry on like this. There isn't anyone here I want in my Marine Corps. You're undisciplined. Unmotivated."* He begins to chew out the faculty.

This Marine recruiter epitomizes what I thought the military was all about. He stood up and took charge and made a big impression on this young seventeen-year-old.

Outside in the hallway, the gunny is folding up his table. A bunch of us swarm to him, eager to learn more about the Marines. I get in line to ask him a question. It's 1958.

I can't remember the question I asked, but we get to talking, and I tell him I'm going to attend St. Anselm College in Manchester, New Hampshire. The Marine recruiter tells me the school offers the Platoon Leaders Class, which is designed for colleges that don't have ROTC.

I sign up and spend my college summers in Quantico, Virginia. When I graduate from St. Anselm in June of 1962, I'm commissioned a Second Lieutenant in the Marine Corps Reserve, and then it's off to Quantico again, this time for six months, for Officer Basic training.

It's peacetime. There's no talk about Vietnam — if asked, I couldn't even tell you where the country is located. The Cuban Missile Crisis is the first indication that I might be going off to war.

My fellow 2nd lieutenants and I watch President Kennedy give his speech on TV. Then everything changes. We're supposed to graduate from basic school next year, in February of 1963. Now it's December of this year, 1962. We commence training 24-7 because our leadership believes we may be going to war with Cuba. We graduate Officer Basic School before Christmas.

In 1965, I'm stationed at the Marine Barracks, Pearl Harbor, as a guard officer. I'm working with a captain from one of the other detachments around the island.

"Here we are, guarding doors and saluting admirals and generals," he says. "There's a war going on, troop morale is low, and we're on security duty."

He comes up with an idea—a program that will allow company grade officers, staff noncommissioned officers (SNCOs), and staff to go to Vietnam on a temporary duty, just a couple of months. They serve in their Military Occupational Specialty, or MOS, and when they return, they'll be able to talk about what's going on in Vietnam to the young Marines on post. He takes it to Fleet Marine Force commanding general Lieutenant General Krulack.

Lieutenant General Krulack signs off on it. Marines who are selected will have to leave during the Christmas holidays.

"Why don't you send me?" I tell the Marine barracks commanding officer. "I'm a bachelor. Let the married guys stay here and be with their families over the holidays. Then, as the program progresses over the year, they'll get a chance to go."

I'm twenty-five, and the Marine barracks commanding officer is the father of six children. He selects me to go to Vietnam. My buddies are all pissed off that they can't join me.

I'm assigned Echo Battery, 2nd Battalion, 12th Marines, a 105-howitzer battery. After four days, I take a Forward Observer

(FO) team out with Hotel Company, 2nd Battalion, 9th Marines. My FO team has a radio operator, a scout sergeant, and a wireman. We're attached to the infantry company to seek out targets in support of the infantry scheme of maneuver.

While we're out on patrol, we get word to return to the rear. Hotel Company, I'm told, is going to be part of an operation that will last five to seven days. We're loaded onto H-34 Choctaw helicopters and dropped off in the heart of the Que Son mountains.

We join up with 2nd Battalion, 7th Marines (2/7). They've been operating in the mountains for several days as part of Operation Harvest Moon, which has been going on for two to three weeks. We're attached to 2/7 and operate with them through the end of the operation. I've been in Vietnam for less than a week, and this is my first operational movement.

Hotel Company's 140 Marines walk in a somewhat canalized line through the mountains. The first few days are quiet. Uneventful. On December 18, we begin to exit the mountains in a battalion movement. Hotel Company is the rear element security. The tail-end Charlies.

I'm pretty sure today is going to be the last day of the operation. As we move and begin to cross a wide-open area, I hear some rounds going off in the distance. I don't think too much of it. Three companies have already passed through the area without incident.

Captain Gormley, the company commander, walks with a map in his hand, a .45-caliber pistol on his hip, the whip antenna of the radio operator behind him. He begins to cross a dike leading into the village of Ký Phú.

And that's when all hell breaks loose. Vast numbers of North Vietnamese (NVA) soldiers ambush us, shooting machine guns, RPGs, and mortar rounds.

I hit the deck. I've never been shot at before—and I'm scared. Anyone who says they're not scared when getting shot at is either lying or smoking dope.

We're taking fire from our left and right and from the rear. The NVA has us pinned down. We're outnumbered ten to one.

The enemy sure knows what the hell they're doing. The NVA are well entrenched and well camouflaged. Disciplined and well trained. They let three companies go through the village of Ký Phú before they triggered the ambush. And then, when they triggered the ambush on our rear element, they also triggered the ambush on the 2/7 companies in the village of Ký Phú.

They've done their homework.

After hitting the deck, I look up from underneath my helmet and see all these young Marines' eyes looking at me. They don't know my name—I've only been with them for four days—but I have a silver bar on my collar. Their eyes are saying, *Okay, Lieutenant, what are we gonna do?*

I'm experiencing fear. And I need to control it. Right now. I need to stand up and lead these men. I need to get over the shock of what's unfolding around me and take action.

My radio operator and I contact the Battalion Fire Direction Center (FDC), explain our situation, and call in a fire mission to take out the trench line to our right.

Our corpsman "Doc" West runs by me and yells, *"The skipper's down."*

I look out on the dike and see, about fifty yards out, a pile of wounded Marines. Captain Gormley is among them. In true communist fashion, the NVA has identified and taken out our leader, Captain Gormley. The communist force believes that if you take out the leadership, the troops will stand around, not knowing what to do. In their forces, officers are the ones who make all command decisions — not non-comms (NCOs). This doesn't happen in US combat units.

I watch "Doc" West run to Captain Gormley. He's shot two or three times and goes down. My scout sergeant, Private First Class McClain, runs out to protect "Doc" and gets shot.

I'm already up and running. Somehow, I reach Captain Gormley without getting hit. I pick him up and carry him back to a covered position. Why I didn't get hit... The good Lord had other things for me to do that day.

Captain Gormley is seriously injured but alive. We talk briefly. I've only known him for three days, but I feel like I've known him forever.

He dies in my arms.

The company commander has just died. And the radio used to talk with battalion command is still out there...

I run back out. The radio operator's dead. I remove the radio from the dead operator and bring the radio back to a secured position and strap it on my back.

"We're being attacked on three sides," I tell the battalion commander. "Captain Gormley is dead. I'm assuming command of the company."

"Where's the XO? The other officers?"

"Sir, they're all fighting right now. I've got the radio. I'm going to start giving orders."

"Make sure everyone knows you're in charge."

I turn the artillery adjustments over to my radio operator, Corporal Iaccunato. I take command and get the company to start taking offensive action against the enemy. I lead a couple of counterattacks against the machine guns that have us pinned down on the right flank. I direct other platoons to do the same on other positions.

High above, I hear jets. I want to use them, but it's overcast, so they can't come down to launch an attack. General Platt, the task force commander, comes on the radio and says, "I've got some armed helicopters. Can you use them?"

"You're damn right I can."

He turns air control over to me. I direct the helicopters against enemy positions. Then I grab an M20 rocket launcher with Willie Peter rounds—white phosphorous rounds used as target markers—and run to a hilltop. I fire the Willie Peter into the trench line. The helicopters come in and hit those targets.

My weapon malfunctions. I use my arms to point at the remaining enemy positions. "You fly down the axis of my arms," I tell the pilots, "and you'll find the target."

They swoop in and do their job.

The enemy is still attempting to close in on us.

Four hours into the battle, the battalion commander, Lieutenant Colonel Leon Utter, who is in the village of Ký Phú, comes on the radio.

"Lieutenant, you've got to come out. We're in one hell of a fight. All the companies are engaged, and I can't come and get you. It's getting dark. So, if you don't come out, you're on your own."

Casualties are mounting rapidly. Ammunition is getting low. I know if we stay the night, there won't be enough body bags in-country to carry us out in the morning.

But how the hell are we going to get to the village? There's five hundred meters of open rice paddies separating us.

I order the engineers to blow down trees to make a landing zone so the helicopters can take out the dead and the wounded. To make ourselves light, I order everyone to throw their packs in a pile and burn them before we commence the breakout.

I radio the helicopter pilot my position. The landing zone, I'm told, is too hot with enemy fire to land. I walk out into the landing zone, the radio strapped on my back.

"Look down here where I'm standing," I tell the pilot. *"If I can stand here, by God you can land here."*

H-34 helicopters fly in and take out the dead and the wounded. "Doc" West has been shot five or six times, but he won't let me evacuate him. He's on morphine, but he's still giving us instructions on how to handle the medical cases that are serious.

When he's the WIA (wounded in action) to be evacuated, I put him on a helicopter. As I'm loading him, he gets shot for the seventh time. It won't be until years later that I find out he lived.

The 2/7 battalion sets up a base of fire in the village. We set one up in our position. The Marines in Ký Phú will help cover us while we break out of the ambush. To get there, we're going to have to cross five hundred meters of fire-swept rice paddy. It's our only way out.

I tell the squad leaders and platoon commanders that we're going to go one squad at a time. "You run as fast as you can. Don't stop. The only time you stop is if a Marine gets shot, you stop and pick them up. We don't leave anyone on the battlefield."

We begin to break out.

My radio operator and I are the last ones to leave. We link up with 2/7 in the village of Ký Phú and fight all night.

Morning comes. The enemy is gone. They've also picked up their wounded and their dead and pulled out.

I'm shocked.

I knew we caught the enemy off guard. They didn't think we'd do something bold like our breakout. You've gotta be innovative on the battlefield. And take advantage and do what you got to do. War is horrifying. It's not glorifying. There was no future in us staying there overnight.

That evening, I'm sitting with the company gunny going over the list of those who were wounded and evacuated, and those that were killed. I think this was my realization of what combat and leadership and command is all about.

I break down. I'll admit it. It got to me.

And then I wonder if some of those who got wounded did so because of decisions I made. My biggest concern was that I'd made the right decision at the right time for the right reason. And that it was a good decision. I didn't know the next day the battle would be all over.

We form up the next morning and continue to move to Route 1 and the end of Operation Harvest Moon. Hotel Company boards trucks to return north to Da Nang, and 2/7 boards trucks to go south to Chu Lai.

Later that night, I'm sleeping underneath a truck when someone pulls on my boot. I come up swinging.

It's a lieutenant colonel. "Settle down now," he says. "I need you to come down to the command group. We've got some questions. I think you're going to get a Sunday school medal for your actions yesterday."

He takes me down to meet a general officer and his command element. I'm asked to explain the battle as I recall it. I go back to sleep, get up in the morning, load up Hotel Company, and start to head north. Ninth Marines sends down an infantry company commander to relieve me and take over Hotel 2/9.

When I arrive back in Echo Battery position, I get cleaned up and taken to the hospital. I sleep for twenty-four hours.

Two days later, I'm grabbing a cup of coffee when the battery commander comes into the mess tent.

"Did you get the word?" he asks.

"What word?"

"Well, we got the call last night that General Walt, the 3rd Marine division commander, has recommended you for the Medal of Honor."

I drop my coffee cup. Shit. I'm just happy to be alive.

Two months later, my temporary duty is up. I go back to Marine Barracks, Pearl Harbor, and assume my duties as a guard officer. I get a call on Valentine's Day that I'm to be decorated with the Medal of Honor on the twenty-seventh of February at the White House.

Three years later, I receive orders to report to Headquarters Marine Corps in Washington, DC, to serve as a general's aide. I've gone from serving as an artilleryman to working with General Walt.

"Most aides don't last a year," he tells me. "If you can last a year with me, you get to go anyplace you want in the Marine Corps."

Fourteen months later, I remind the general that I lasted over a year with him and that it's time for orders.

"Where do you want to go?" he asks.

"Vietnam." I'm a bachelor, there's a war going on, and I'm a professional Marine, so that's where I should be.

"You can't go back to Vietnam," he says. "You've got the Medal of Honor."

"General, you told me if I lasted with you a year, I could go anyplace that I wanted to, and I want to go to Vietnam."

I'm the first Vietnam Medal of Honor recipient to return to Vietnam.

It's never been about me.... It's about the team. I've worn this medal for fifty-eight years now in honor of the great Marines and the phenomenal Navy corpsmen that I got to lead on the field of battle that day in 1965.

I helped save about 130 Marines. That's more rewarding than the medal.

During one of my follow-on tours, I'm a battalion commander

at Parris Island. I'm involved in turning unorganized civilians into low-crawling, hard-charging United States Marines.

To become one, you've got to make a commitment because it is a commitment. Set your goals high and reach out to get them. It's going to be tough, but don't quit. Don't bring the word *failure* to Parris Island with you because there ain't no such word. Never say I can't, never say it's too hard.

And if you try something that doesn't work, you don't quit. You stop, back up, reevaluate, and take another avenue of approach. If you're going to be a bear, be a grizzly bear. If you ain't, don't even go the first time.

Matthew O. Williams

Sergeant, US Army
Conflict/Era: War on Terrorism (Afghanistan)
Action Date: April 6, 2008
Medal of Honor

I'm a freshly graduated, brand-new Green Beret, and next thing I know I'm in Afghanistan. It's the end of October, right after my twenty-fourth birthday. I'm about to become a member of ODA 336, 3rd Special Forces.

My father instilled hard work, integrity, selfless service, and other values in my upbringing. I've always had a willingness to serve, but the military wasn't something I thought much about until 9/11. After that, I went about my business, learning as much as I could about the different branches of the military.

When I graduated college in 2005, our nation was at war. I settled on the US Army's 18 XRAY program, which allowed me to go to Special Forces assessment and selection right off the street. I liked the Green Berets, their mission and motto, *De oppresso liber.* It means "free the oppressed." Giving people the opportunity to fight for themselves is very important.

I sign in and meet my team. These guys have been training together for six months, and there aren't a lot of new guys in the room. It's scary enough meeting these guys for the first time, wondering if I'm going to fit in, how I can be a successful part of the team.

"Get your stuff together," Team Sergeant Scott Ford tells me. "We're leaving."

These guys, I discover, are great leaders and mentors. Staff sergeant and medic Ronald Shurer takes me under his wing very quickly. Everyone shows me the ropes.

We're assigned to the first kandak, or operational battalion, of Afghani commandos. Our military has provided years of training comparable to that received by a Ranger battalion. Now we're going to work with them.

My first mission will be village clearance with some other Operational Detachment Alpha (ODA) personnel. I load up in the helicopter in Bagram and fly out. As we're hovering over the target, everything around me completely browned out, I look down at the dirt, half expecting the bad guys to rise up and attack.

And that's when it hits me. I'm in Afghanistan, and this shit is for real.

We're able to maneuver to our target building, do our charges, and begin our assault. We spent the next four days clearing the village and get in a couple of little scrapes here and there. I learn a ton.

The next six months are fast-paced. Either we're training or prepped to go on missions. It's awesome. I'm doing exactly what I want to be doing, what I was trained to do.

On April 6, 2008, we wake up early, before dawn. We've been charged with doing a capture/kill mission for a high-value commander of Hezb-E-Islami Gulbuddin (HIG), another terrorist faction in Afghanistan/Pakistan. He's operating out of a mountaintop village in the Shok Valley on the Afghan-Pakistan border.

It will be a normal raid, the kind we've successfully executed several times. We'll be flying multiple helicopters filled with commandos. We start doing our own internal team checks, from gear to equipment.

I'm in charge of the Afghan commando heavy weapons squad for this operation. I make sure everyone has their machine guns and ammo, and as we load up, we're told to stand down due to weather. We're on hold, which always kills momentum a little bit.

The weather clears a few hours later. Our whole schedule has been blown. As we fly in Chinooks, Scott, our team sergeant, stands at the back of the bird and gives a little pep talk to get us remotivated.

The helicopters come to a hover over the valley. It turns out that the birds can't land on the river bottom because the rock and boulders will puncture the helicopter bellies or cause a crash.

The ramp lowers. When it's my turn, I look down and see we're about ten feet off the deck. Waiting below is an icy, semi-dry river. I look at the crew chief, my expression saying *really*?

Yeah, jump down, his expression says.

I jump and join the others.

We dust ourselves off and get everything back in order. It's midmorning. Time to head up to the village.

Our lead assault element consists of a couple of Americans and a bunch of Afghani commandos. Our C2, or command and control element, is led by our captain, Kyle Walton; our intelligence sergeant, Luis Morales; Dillon Behr, our communications sergeant; Mike Carter, our combat cameraman; and Zach Rhyner, our JTAC, or Joint Terminal Attack Controller, the person who coordinates close air support and other offensive air ops.

I'm in the trail element along with my team sergeant and Staff Sergeant Ron Shurer, who is our one and only 18 Delta medic. We have our commandos and another 18 Echo, Ryan Wallen, and Seth Howard, another 18 Bravo.

The lead element makes it to the village outskirts. The trail element is about halfway up this terraced, stair-step mountainside when everything erupts. Rifle and machine gun fire and RPGs are coming in from what seems like all directions. It takes us a while to figure out that's not the case. We keep returning fire as best we can.

The first call comes over the radio. "We have a wounded American."

One of the guys in our C2 element has been shot in the hip, near the femoral artery, which is a big problem. Luis Morales goes to render aid and is shot. Now we have two wounded guys who are pinned down halfway up the mountain.

Team Sergeant Scott yells out to our medic. Ron tells us he is going up.

I don't know why — maybe I'm suffering from a little bit of

FOMO — but I holler up to Scott, "I'll come with you. I'll bring my heavy weapons guys to help out."

"Let's do it," Scott hollers back.

We run across the valley floor and then make our way through a river. The water is ice-cold and waist-deep. We fight our way up the mountain and make it to the C2 element. Dillon Behr has been shot in the hip, Luis Morales in the ankle and a couple of other places. Our interpreter, Edris Khan, has been killed. I order my commandos to provide suppressive fire.

Ron grabs our combat cameraman, and as the two of them get to work, rendering aid, Kyle Walton, our detachment commander, myself, and Scott Ford begin to look around and take stock of our situation.

There's this weird lull. Scott and I start talking to two guys who are making their way back from the village. They're probably one hundred meters from us. We have a candid conversation about what we're going to do: maneuver the two wounded men down the mountain and get them somewhere safe. We'll establish a foothold in the village and then systematically go to work.

I'm moving down the mountain to prepare to move the wounded men when I find out a sniper has hit Scott in the chest plate and arm. Another soldier, John Walding, is shot in the shin. The round amputates his lower leg on impact.

Now we have four casualties.

I climb back up the mountain, back to their location.

The enemy's gunfire has become so intense and relentless that for their protection, Scott and John have been moved to the ledge of a cliff. Scott has a tourniquet on his left arm, a semblance of a bandage, and he can move a little bit. His chest plate worked, preventing any trauma to his thorax.

I pick Scott up, help him down one level, and hand him off to Staff Sergeant. Seth Howard, the Team Weapon Sergeant and a trained sniper.

"Take him down," I say. "Take him to that covered mud structure at the bottom of the mountain. We'll make that our casualty collection point. It's at least covered and semi-protected."

Seth readjusts Scott's bandages, and as he takes Scott down to the mud structure, our JTAC is continuously dropping bombs on the village, to little effect. The enemy is dug in. The air strikes aren't slowing their attack.

The rest of us are hunkered down, doing everything we can to protect our medic so he can work on our four casualties. The best thing we can do now is consolidate our wounded somewhere safe and then go exfil and try another day.

Seth returns. He and I are colocated one step below the casualties. We're returning fire, but we don't have good visibility on the enemy, and we can't get the guys above us on the radio. I'm worried they're either wounded or dead.

Seth and I decide that we're not going to leave them there, no matter the circumstance. Because of the heavy fire, there's no way we can climb up the mountain directly. The enemy has it too pinned down. The only way up is to climb laterally, hand over hand, on the cliff face.

We get together with our Afghani commandos and tell them the plan.

"We're going back up. You guys follow us and then you'll get up there and we'll provide suppressive fire."

We take off. When we glance back, no one is following us.

Seth and I climb nearly thirty feet, sideways, to a little rocky outcropping that offers a vantage point. Kyle, our detachment commander, is with his commandos.

Kyle spots us. His SATCOM, or satellite communications, radio is all jacked up, so he's unable to call in strikes. He tosses it to me and says, "Fix our radio."

I'm able to do it because of good team cross-training.

One of our guys has found what's basically a trail/mudslide down the backside of the cliff. Seth starts laying down covering fire with his sniper rifle, to prevent the insurgents from overrunning their position. I join Seth, hoping our combined fire will help everyone, including the wounded and the dead, to get back down the mountainside.

The firefight lasts several hours. During that time, our JTAC calls in a danger-close two-thousand-pound bomb. It explodes seventy meters away from our position. We're separated from the blast vertically, and the explosion just blacks out the sky. It helps us move down the valley with cover.

Ron continues to treat four casualties with one aid bag, which is a miracle in and of itself. Our JTAC continues to drop bombs on the village. As the medevac helicopters fly in, they're getting shot up. One of the pilots is wounded, but we're able to get our casualties and our deceased interpreter out of the area.

We consolidate our forces, make sure we have everybody and everything, and we're able to exfil.

The team is awarded ten Silver Stars and one Air Force Cross. What still stands out for me that day, what I've learned, is that when everything hits the fan and it's game on, if you have trust and real camaraderie with a really well-trained team, your capabilities alone and together are pretty astounding. I know I'm going to constantly seek out this level of camaraderie and team building throughout the rest of my military career.

Ten years later, Ron Shurer, our medic, and I get the call that our Silver Stars have been upgraded to the Medal of Honor.

I'm over in 3rd Group and still a team sergeant on an ODA when I receive the award. My community is small and tight-knit and quiet. If somebody wants to ask questions or talk about the medal, I'm more than happy to discuss it, but more often than not, they don't.

Everybody understands I'm just a guy. I never really wanted to do anything other than be a Green Beret. I want to continue to serve.

I've made it a point to just be "Matt" and separate myself from the award. Now I find myself in leadership positions that are extraordinarily fulfilling because I'm able to give back to the younger guys. Do I highlight the heritage and history of the regiment by using my story to engage with these kids so they can learn more about the history? Or do I stay hidden?

Frankly, I still haven't decided.

Donna Barbisch

US Army, Major General, Retired
Conflict/Era: Vietnam War

Every night, Walter Cronkite brings the Vietnam War into my living room.

I'm nineteen and just finishing my second year of nursing school. Options for middle-class young women like me are fixed: nurse, teacher, secretary, or hairdresser. I like medicine and the flexibility of careers in the medical profession, so I choose nursing.

The Vietnam War is raging. Every night, Walter Cronkite gives the latest casualty report. Tonight, the details feel personal. I think of the men serving over in Vietnam. Most of them were drafted. Whether they think the war is right or wrong, whether they did or didn't want to go to war, they're out there doing their jobs, serving our country.

The war has polarized Americans, but we *must* take care of our guys. They need our support — and our help. My help. I'm still a nursing student, but I decide to see an Army recruiter and volunteer to go to Vietnam. I didn't know where it was, I look it up in our encyclopedia. I have this overwhelming sense that taking care of our troops is my higher calling. I'm going to do some good.

The Army assigns me the rank of private and pays me ninety-one dollars a month. When I pass my nursing boards, I'm commissioned as a second lieutenant and sent to San Antonio, Texas, to attend the Army Medical Department, Officer Basic Course.

It is January 1969. Some men resent my rank. Some deliberately cross the street so they don't have to salute a woman. We march. The band plays "Thank Heaven for Little Girls."

While at Fort Sam, I learn the workings of the Army. Medical personnel are taught the damages caused by the munitions it deploys—weapons and bullets, shrapnel, and booby traps. I've never seen anything like that!

Then I'm sent for field training. In two weeks, I'm taught enemy tactics and techniques, how to read a map and how to survive in the jungle.

The Army is going to take care of me, enable me to do my job, give me all the resources and necessities I need to succeed in military life.

That's how young and naive I am.

I'm twenty-one years old when I deploy in September of 1969, one of two women on an airplane flying from Pittsburgh to Hawaii, then from Hawaii to Tan Son Nhut Air Force Base, outside of Saigon. When we arrive, we have no assignment. The Army doesn't know where to send us.

Wait a minute. Didn't you know we were coming?

The world feels upside down. There's a twelve-hour time difference and, oh, my God, the weather is *hot, hot, hot.*

I spend a few days and nights in hot, sweaty tents. The few women in-theater are considered somewhat of a novelty. *What have I done?*

Finally, I get my assignment. I'll be working at the 91st Evacuation Hospital in Chu Lai, about sixty-five miles south of Da Nang and right on the South China Sea.

I'm ready for this, I keep telling myself. *I'm trained and I'm ready.*

At the start of the Vietnam War, Army medicine was virtually unchanged since World War II: from first treatment by medics in the field, wounded soldiers were sent to the closest aid station. Then to collection and clearing stations, then to hospitals. First a field hospital, then an evacuation hospital, and finally a station or general hospital for more definitive care. The time elapsing often proves fatal for the soldier. Emergency nursing was not even a specialty back then. The specialty was developed because of Vietnam.

In 1962, the first Air Ambulance is assigned to Vietnam and adopts the call sign "Dustoff." They pick up wounded soldiers from the battlefield and in five or ten minutes deliver them to the nearest hospital. While the first use of helicopters to evacuate patients occurred in WWII, the Army's medical system refined the process and redesigned helicopters to accommodate litters and the walking wounded, as we call them, in Vietnam. We get patients within minutes of their injury, saving a lot of lives.

AMERICAN HEROES

The TV show M*A*S*H (Mobile Army Surgical Hospital), while set in Korea, was really about the Vietnam War. Tents and round, corrugated metal Quonset huts connected by wooden structures are the template for our hospitals.

The 91st Evacuation Hospital has, roughly, a 325-bed capacity, thirty-five nurses, and twenty-seven surgeons and doctors. I'm assigned to R&E—Receiving and Emergency. Our radio has a live feed to the dispatchers speaking with the helicopter pilots.

The radio announces patients are incoming with an ETA of five minutes. Immediately we assess, stop the bleeding as much as possible, start huge IVs into their collapsed veins, and send them to surgery. I am part of an amazing team. It was like drinking from a fire hose.

A week after arriving, it is my first night alone in R&E. It starts out quiet. I am in charge with two corpsmen on duty. Then the radio starts to crackle. We have incoming casualties. The corpsmen leave to wake up the medical team. I call for blood and alert the OR.

I can see tracer rounds going off but can't hear the siren alerting us to incoming mortar rounds and signaling us to take shelter in the Quonset hut covered with sandbags. This past June, a nurse named Sharon Lane died right here, at this hospital, from incoming fire. Back then the buildings weren't reinforced with sandbags.

I don't have a fear of death. I know I'm in harm's way, but I've always been a little pragmatic—walking across the street puts you in harm's way. Though to some people it reads as cold, saving lives means separating what I have to do and dealing with it. I can't let myself get too emotional because I won't be able to do my job if my empathy overcomes my ability to act. I've been pretty good at doing that most of my life.

I hear the *whoop-whoop-whoop* of helicopter blades growing louder. Closer. *Please let the team get here before the patients.*

Our systems are in place. IVs are hung and ready to go. I double-check the supplies on my uniform; my hand brushes past my scissors, IV tape, straight razor as I watch the helicopter approach. Dust is everywhere.

The full team is here. Doctors, nurses, and corpsmen. The chaplain is also here, as well as a corpsman who works in Graves Registration. GR men are responsible for processing bodies and returning them to their families for burial.

The helicopter lands, and we're off and running.

Dozens of litters are removed from the helicopters. The wounded are covered in blood. Brains and body parts are unrecognizable. Horrific injuries. Some with bones penetrating their skin. They are in shock.

We don't have time to check a patient's blood type before transfusion, so the blood bank stocks the blood coolers with low-titer group O whole blood from O positive donors. It contains red blood cells and plasma with low-level antibodies, a formula that expands blood volume and is relatively safe for a patient with any blood type to receive.

IVs are set up next to litter stands resembling sawhorses. We use a 16-gauge IV. They are hard to start when veins are collapsed from shock but critical to replacing fluids and blood. We start two on each patient. After the patient is prepped, he's moved to the pre-op area or straight to the OR.

The whole process takes five to seven minutes, which is pretty remarkable. But what I'm wholly unprepared for is the reality that triaging casualties comes down to making cold, hard life-and-death calculations in a resource-constrained environment.

In mass casualties, it is all about saving the greatest number of patients. It is a hard and emotionally taxing numbers-driven guide. Those that are critically injured, requiring overwhelming and currently unavailable resources (docs, nurses, supplies) are put into the "expectant" category, meaning we move them to a quiet area and make them as comfortable as possible while we treat the largest number of patients who have a good chance of surviving. If the severely injured patients are still alive after we stabilize the others, we get to work on them.

One soldier lost a leg—we replaced enough fluid and he's awake. He grabs my collar and says, "You've got to let me die. I can't go home like this."

Over and over, soldiers arrive with missing limbs and make the same demand. Back home, prosthetics are primitive, and the Army lacks the medical and psychological capability to make these soldiers whole, physically or mentally.

I start an IV on another patient, then turn to the patient who's been hit in the chest with an M16 round. I grab my scissors. We can't assess the damage until his clothing is fully cut away. The doctor working next to me is new. He starts to explore the injury before I expose the patient's body and his fingers graze my scissors.

"Oh, I didn't mean to cut you, Doc. Sorry."

The entrance wound is the size of my little finger, but the exit wound is nearly ten inches wide. We manage to stop the bleeding and resuscitate the patient. Bringing him back to life is the reason I am here. War is awful but I am making a difference. It is one of my most rewarding experiences.

But when I go check on him, he isn't in the recovery room or the ICU. I go to the OR, where I'm told he died during surgery. *What? Why? How? We got him back!* This time I cry.

His lungs shut down. What we don't yet know is that when the lungs collapse, tiny air sacs called alveoli release a sticky substance called surfactant. When this happens, the patient can't oxygenate. A lot of people die before we identify what will become known as "shock lung."

When patients in the expectant category don't make it, we "process their remains," a technical term for sorting through what they have on their body when they die. As I empty their pockets, I often connect with the pictures of families or the notes they carried. These things speak volumes about their lives.

This experience does not affect me.

At least that's what I keep telling myself.

I call home. The first time, I use the radio. When I say, "Hello, Mother. Over," she becomes befuddled. It is awkward to have the radio operator switch the line back and forth, but it is good to hear her voice. Next month I go to the USO and stand in line for the two hours it takes to use the phone. I call. My mother is overwhelmed and tells me the news reports a lot of fighting south of Da Nang. She cries as I try to assure her and tell her the entire country is south of Da Nang.

If the sound of my distant voice upsets her this much, it's not helpful. For either of us. I stop calling.

The emergency room is an adrenaline rush. I work twelve-hour shifts, six days a week—unless we're extremely busy. Then I work seven days a week. There are periods when I'm working continuously for up to twenty-four, twenty-six hours.

A lot of medical staff suffer burnout, but the intensity fits my personality. Still, the magnitude of my job and the horrific injuries I encounter daily and at high volume—it's often overwhelming.

After six months in R&E, the chief nurse decides I need to rotate. I don't want to go. It is for my own good, he says. He assigns me to the medical ward. It's the exact opposite of R&E. Slow pace, noncombat casualties, long, drawn-out days, no adrenaline.

Oh, my God, shoot me. I need to get out of here.

In 1969 Vietnam, malaria is rampant. Processing medical tests to analyze a blood culture confirming the diagnosis takes time, three to four days. While patients wait, they're running fevers of 105 and 106. We dose them with aspirin and Tylenol. We pack them in ice to bring their fevers down. We make our own ice. Funny, a lot of guys love coming to the hospital because it's the only place in Vietnam that has ice. Some things are just weird like that.

Antibiotics are in constant demand. Sexually transmitted diseases are a huge problem. About 25 percent of the GI population in Vietnam are infected. Many strains are antibiotic resistant and can't be cleared up without hospital-given IV antibiotics. The medical ward has a lot of STD patients. My naive values are shattered.

In addition to taking care of our soldiers at the 91st, we take care of the Vietnamese population. After two months on the medical ward, I volunteer to serve on the Vietnamese ward. It is not a popular assignment, but at least it is more interesting to me. The patients—civilians, military, North Vietnamese and Vietcong and what we call the detainees, the men, women, and even children who have not been identified and might be good or bad. We have MPs on the ward to watch over the POWs and detainees.

We don't have any cultural training or professional interpreters. A lot of the Vietnamese patients think we're going to hurt them. One patient with abdominal injuries isn't allowed to eat or drink and doesn't understand why. He bites into his IV so he can drink the fluid. There are many loving and kind people hurt by this war. With a broken femur from a shrapnel injury, the father of one baby crawls toward a crib to comfort his infant son.

A year later, September 6, 1970, my Vietnam service is over. I fly home to Pittsburgh. We're not allowed to wear the uniform in public so I change into my musty, wrinkled civilian clothes. Serving our country carries a stigma, even within families. One servicewoman I know told me the day she got home, her mother said, "I'm glad you're back. You did it. We don't ever want to hear about it again."

Some people throw their uniforms away. I keep mine.

AMERICAN HEROES

I change planes in Chicago. A guy comes over to me and says, "You're coming back from Vietnam, aren't you?"

Uh-oh. A lot of soldiers are being targeted for the government's role in the war. *Now what?*

"The day you flew to Vietnam," he says, "I was on the plane with you. I remember you because there were only two women on that flight."

We strike up a conversation. I'm lonely. I miss my war family. It is comforting to be with someone who understands. I tell him I'm taking the bus from the airport to downtown Pittsburgh. There, I'll change buses and eventually get to my mother's house. I haven't told her that I'm coming home.

My new friend and his wife live not too far from my mother. He insists on driving me home. When I arrive, I find the house empty. *What a relief.* When my mother sees me, she'll cry, and I can't deal with that. I'm too emotionally raw.

I call a few friends to let them know I'm home. They come by and talk about what movies are playing and what they're going to have for dinner—all these mundane daily activities. I listen to them and all I keep thinking is, *I don't care about movies or what's on TV or what I eat. People are dying.*

I can't cope with the small talk.

And then there are people I meet who don't believe I served in Vietnam. "No, no, you couldn't have been there," they tell me. "We didn't send women."

My next duty station is Fort Devens outside Boston, Massachusetts. I call up my detailer and ask if I can report early. He says yes.

I get in my car and leave. It is good to be alone.

* * *

I went to Vietnam to make a difference for soldiers who served our country and needed medical support. Only eleven thousand women served in Vietnam. Ninety percent were nurses. We were all volunteers.

I've been back in "the World" for six months. My active-duty commitment is over. *What should I do?* The Army is in my blood, but I'm lost. I move from active duty to the Army Reserve.

I become a nurse anesthetist and get married. I go back to school to attain my bachelor's degree. Two kids, then divorce. Then back to school for a master's degree in public health. Within my military role, I become the commander of a Mobile Army Surgical Hospital (MASH), a first for a nurse. Then on to a doctoral degree. My dissertation — "Identification of Barriers to the Use of Department of Defense Medical Assets in Support of Federal, State, and Local Authorities to Mitigate the Consequences of Domestic Bioterrorism" — is published in early 2000.

My colleagues think I'm crazy. The Berlin Wall is down and the USSR is dissolved. Terrorism, they say, is a ruse for the military funding. They tell me the country will never use military assets domestically. We only fight overseas.

Planning, I tell them, is inexpensive. We must ask *"what if?"*

In August of 2001, a month before 9/11, I present a lecture to the Army War College on the military role in domestic response and the need for "surge capacity," a term I have to define because the concept of mass casualties in the hundreds or thousands is unfathomable in the US. My understanding grew from my experience in Vietnam.

I become the first woman in the Army Reserve to rise from nurse to major general. We as women were not groomed for leadership roles. Women have a tendency to be a little bit more collaborative than our male counterparts. Collaboration works to gain insight, build value and support. For a leader, it builds followership. When the buck stops here you need to be decisive and provide direction. We need followers to make it all work.

I only talked about my experiences in Vietnam to people who were there. We would sit together and peel the labels off our beer bottles as we shared our stories. I hear the choppers and think about the incoming casualties. I wonder how those soldiers are doing today.

I do some work for Wounded Warriors and welcome home soldiers and their families. One day I watch two servicemen discussing their new titanium legs.

"Which one did you get?" a soldier asks. "I'm a runner, and I want to use a leg that helps me run faster."

I think about all those Vietnam soldiers who grabbed me and told me they wanted to die so they didn't have to go home without a leg. I cry! Now we have a system in place that can make these brave men and women whole again...happy they have it better. Those soldiers in Vietnam didn't have the option.

Nearly three million men served in Vietnam. More than fifty-eight thousand lost their lives. Many came home with their lives changed forever. Most of those who returned—not all but most of them—threw their uniforms away and never talked about it again. Back then, we blamed soldiers for the politics of war. We didn't thank them. We didn't recognize their service or their sacrifice.

It's been fifty years, and we haven't done enough to help these men and women. We haven't properly recognized what they did for our country. Fifty years and they still aren't truly appreciated. Knowing that, the little we do is almost adding insult to injury.

I carry these little brass-colored "Vietnam War Service" pins. When I see someone wearing a Vietnam hat, I go over and ask, "Did you serve in Vietnam?"

"A little," they'll say, somewhat hesitant.

"I was there with you and I appreciate your service," I say, and give them a pin. They smile and look up, often with a tear in their eye. It should be so much more. It needs to be.

David G. Bellavia

Staff Sergeant, US Army
Conflict/Era: War on Terrorism (Iraq)
Action Date: November 10, 2004
Medal of Honor

Today is my birthday. I'm twenty-nine.

It's November 10, 2004. I'm a staff sergeant with 2nd Battalion, 2nd Infantry Regiment, stationed in Fallujah, Iraq. I'm near the end of a thirty-six-month "all others" tour away from my family, currently deployed to Iraq in support of Operation Iraqi Freedom.

Fallujah had been abandoned for six months when we arrived in the late fall of 2004. During that time, four thousand to six thousand enemy insurgents have entrenched, preparing their defenses for our arrival.

Bodies are all over the street, festering bacteria. Within a matter of days of our arrival, we've all suffered strep throat, fevers, and diarrhea. It's horrible.

We engage in close-quarters combat, within a deadly two-foot radius. The enemy is a mix of highly skilled professionals and amateurs who fight with passion. We never know what we're going to encounter.

I'm not bothered by fear. I'm fueled by it.

On November 10, 2004, we take part in "Operation Phantom Fury." We're informed that six or more insurgents have taken refuge inside a block of twelve homes. Civilian Michael "Mick" Ware, an Australian journalist for *Time* magazine who's been embedded with US forces in Iraq since 2003, is alongside my unit as we're tasked with searching and clearing the homes—and capturing or killing the insurgents. A Bradley fighting vehicle (BFV) will accompany us to provide fire support.

We set out. It's 1:45 in the morning, so dark we can barely see. None of us have had any sleep for the last forty-eight hours.

We dismount the BFV and make our way through the darkness, blinking ourselves awake. Our yawns give way to adrenaline as we hit door after door.

We freeze in place when we sense the enemy. We engage in extreme close quarters, constantly draining and dumping adrenaline. Instead of being tired from lack of sleep, we're now all suffering from adrenaline hangovers from clearing over three hundred rooms. They're all empty except for one that shows signs of recent life.

We methodically clear buildings one through nine without incident. We don't find any insurgents, but we do find a lot of weapons, including AK-47s, rockets, and grenade launchers. We also find a lot of needles and drugs. The insurgents are using adrenaline and amphetamines, even opioids, to provide stamina to fight and to continue fighting even after being hit by multiple rounds.

It's like we're battling zombies.

We approach the next compound—house number ten—and immediately clear the first room. It's pitch-black inside. Our rifle lights show us trash and debris, conditions of abandonment. But through the dust and grime, I can smell that someone's been living here.

A single doorway leads into a rectangular-shaped space. We move into the area. Another doorway to our immediate left leads to a room with a staircase, and we take on intense fire from two insurgents entrenched underneath the stairs and behind a half-ton concrete Jersey barrier. The rounds are coming from an AK-47 and a belt-fed PKM machine gun made by the Soviets.

The enemy created a kill zone and we've stepped right into it. We drop to the floor and scramble for cover as a hailstorm of bullets tears through the walls above our heads. We can't escape the house because more insurgents have positioned themselves by an adjacent window, and they're keeping us pinned inside by splitting our outside security and fire support from us in the building.

We have gunfire coming at us from two directions as our machine gunners fire back into the kitchen insurgents.

Inside, we have the same idea. Suddenly, I can no longer return fire.

I check my weapon. A round hit the magazine, rendering my rifle useless.

I couldn't have asked for worse luck.

Some of my men in the other room are wounded. Two have bloody face lacerations from glass and metal shards and another has been hit in the stomach under his vest. They're trapped and are going to die unless something is done.

I grab the SAW, the M249 Squad Automatic Weapon, a machine gun with a basic load of six hundred rounds of linked ammo.

I go through the doorway, stand in the fatal funnel, and start firing.

The insurgents can't compete with the SAW. They duck behind the barrier as my rounds rip through walls and devour wood and tear chunks of concrete from the barrier. As my rounds bite into the concrete, the barrier's Styrofoam innards (designed to contain bomb blasts) turn into white flakes that tumble through air. I keep firing at a cyclical rate, allowing my platoon a chance to escape.

By the time the SAW clunks empty, everyone in my platoon has left the house.

Now I've got to get out of here.

I bolt outside and run away, hearing and feeling rounds whisking past me, seeing them tear up the dirt. The insurgents are alive, still in the compound.

I had the enemy—I *saw* the enemy. And I broke contact.

It's one of the lowest moments in my life.

I regroup with my men. A BFV arrives and fires 25 mm rounds into the compound. The insurgents are still occupying the building, shooting at us from the windows and the roof.

"I wanna go in there and go after 'em," I tell my squad.

I take a soldier and place two SAW gunners in the courtyard. I'm not bringing all these men back inside to be shot. Staff Sergeant Scott Lawson, from Michigan, is armed with only a 9 mm handgun, and I entered the house with Mick Ware, an unarmed reporter, following closely behind. I tell Ware to run outside and leave the area. He agrees and then keeps following. Clearly, he isn't concerned for his safety—and in a strange way, it gives me extreme confidence in whatever idea I might employ once I get closer to the enemy positions.

As we reenter the house, I look around at how the BFV 25 mm cannon rearranged the floor plan on the house. It's almost unrecognizable. I hear two men whispering on the other side of the wall.

Through a reflection off a broken mirror, I can spot one of the insurgents loading an RPG launcher. My mind starts racing. I think of possible plans, but my mind is so exhausted I can't come up with any good ideas.

If that rocket fires, we all lose.

I have an idea, one that has a major potential of a horrible ending.

There's no choice. I lower my head and storm into the room to engage the insurgents.

Their return fire is fierce. I'm forced to fall back. When I reengage, I tell Ware to stay in the first room. My partner, Staff Sergeant Lawson, turns the corner and starts putting rounds down.

The insurgent with the RPG doesn't have anywhere to go.

He's trapped. He goes down as a second insurgent runs through the house, firing wildly.

We exchange gunfire. I hit him several times, but he doesn't go down. He's still on his feet and he's still running and firing because he's jacked up on amphetamines and adrenaline and God only knows what else. These drugs keep them in the fight even after they've been seriously wounded.

The insurgent flees into the kitchen. My ears are ringing from close machine gun fire. Lawson is down to one magazine, and his left arm is wet and completely limp. He promises to tough it out. I get on the radio and give a situation report. Then I dump the radio. I don't want to give my position away to the enemy.

As I move through the house, some places are flooded with ankle-deep water from destroyed plumbing during our earlier BFV barrage. It's slimy and has a pungent, fishy, toxic odor. Finding footing is difficult, and the smell is overpowering.

An insurgent moves down the stairs and starts firing at me. I find cover against the wall. As I'm thinking about what to do next, I hear, behind the gunfire, screams coming from somewhere upstairs.

The insurgent I saw flee inside the kitchen hasn't been mortally wounded. I see him moving toward a bedroom doorway. He blindly fires his AK-47, but I've got the jump on him. I put him down and move farther into a dark room.

It's a bedroom.

The insurgent on the stairs starts firing again—and now he's changing position, moving toward me.

The bedroom has a wardrobe standing in more puddles of

noxious water. There's a gap in the room's framing, this little bit of space between the concrete and doorframe. I wait. Watch.

I catch sight of the insurgent from the stairs. With my infrared laser, I'm able to fire with precision through the gap.

Threat eliminated.

As my brain processes my precarious situation, I try to lower my heart rate by taking deep breaths—which isn't easy. The room reeks of death.

As I move down a length of wall, there's a sudden rocking near me in the dark. My spine stiffens. I ready my rifle, my heartbeat back to full gallop.

There's a crash followed by what sounds like movement from inside the wardrobe. The door suddenly flies open and a guy jumps out armed with a snub-nosed AK-47 tucked under his armpit.

The weapon comes up, and as he begins firing, he falls, pulling the wardrobe down with him. Miraculously, the wardrobe lands with both doors open, providing me with a decent area for cover.

The insurgent scrambles to his feet, firing blindly, and immediately trips on a mattress. He loses his balance. I break cover and put rounds down on him.

Right before he falls, he fires a burst in my direction. The rounds narrowly miss me. Next thing I see is the wounded insurgent running out of the doorway, as if he doesn't have a scratch on him. He's like the enemy who fled into the kitchen, seemingly jacked up on something.

I break cover to follow and end up slipping on his blood. As

I tumble forward and hit the ground, enemy rounds tear into the area where I had been standing just seconds ago.

I follow the blood trail up the stairs and realize I'm almost out of ammunition.

The bad guy is inside the bedroom to my left. I can hear him. I toss a grenade into the room.

The explosion goes off.

He's still alive — I can hear him yelling. I assume he's giving away my position.

As the bedroom fills with smoke and flames, I hear another insurgent yelling back to the other one. The voice is coming above me. Is there a third floor? The men are discussing me. I'm sure of it.

The only thing I can see is a door. I move to it.

Looking through the smoke, I discover that the wounded insurgent is standing.

He rakes the butt of his AK-47 across my face.

I hit him with my rifle, as we fight each other with any object we can reach. We move at each other. It's panicked fighting, the two of us unsure at how we found ourselves in this situation. The room seems to be filling with smoke as we roll around the wet floor. I smell blood. I quickly use my helmet to hit him repeatedly. He sinks his teeth into my wrist like a feral dog.

I scream and hit him with the ballistic plate from inside the vest of my carrier. One hand on his hair, ramming his head in the ceramic body armor. Slowly, he stops resisting. Just when I let my guard down, he digs his fingernails deeper and tries to break free.

I remember my knife. I remove it from my cargo pocket and sink the blade into the insurgent's collarbone and neck. I hold it there until I'm absolutely, 100 percent sure he's dead.

I'm wounded and beaten up. Stressed out. My eyes are burning from the smoke, and I can barely stand, let alone walk. I'm physically and mentally exhausted.

I stumble outside. The door opens to the second-story porch sitting underneath a third-story rooftop with stairs. The insurgent I just killed was yelling up and from that direction and I heard somebody shout back. As long as I keep my eyes on those stairs, I can wait this out.

Seems like the perfect time to smoke a cigarette. My guys can take out the fifth bad guy because I'm done. Spent. I've got nothing left.

I'm smoking when the wounded insurgent jumps from the third- to the second-floor roof. He lands on the ground, staggers.

I'm not wearing my helmet and my weapon is resting a few feet away. He's got the jump on me.

In a daze, I go grab my weapon. He's already off and running. I fire at him, hitting his legs and lower back. He drops to a lower floor but is still alive. While I'm looking for more ammo, he falls off the roof into the garden below.

Below me, I hear machine gun fire as members of my platoon enter the house. We don't get the chance to find the sixth insurgent, if there was one — my company commander has called in a bomb.

Now it's time to regroup and prepare for our next crazy fight.

* * *

"Into the Hot Zone," the article Mick Ware writes about that night, is *Time* magazine's cover story less than two weeks later.

I earn a Silver Star. But all I hear for ten years is, "Bullshit. I don't believe it. That didn't happen."

Then I get a call from the military paper *Stars and Stripes*. "Hey, you're nominated for the Medal of Honor, did you know that? I hear there's a videotape. Do you have a comment?"

I'm immediately on the defensive. No one who's served in Iraq has received the US Armed Forces' highest military decoration, except posthumously. "What's on the tape? How did you find out?"

The Army's trying to tell me that I'm getting the Medal of Honor, and I'm acting like they're trying to put me in jail.

Turns out Ware sold a documentary to HBO. He filmed the entire firefight. Honestly, Ware's anti-war and pro-freedom for the press to tell the truth, but he's got the biggest balls of anyone I've ever met in my life. He was right there the whole time. Because of Michael Ware, everything is corroborated. He was recording the fight the entire time.

They name me the first living recipient to earn the congressional Medal of Honor for bravery in the Iraq War.

A coworker of mine reads about the award. "Hey, some guy with your name is getting the medal of honor. Isn't that weird? How many David Bellavias are out there?"

"I know, right? It's so weird," I say.

It's surreal and unnatural to get credit when you've lived

your entire adult life to be about the team. It's never about the individual. I'm not there to celebrate me.

I decide to be the first guy to bring his entire unit to the ceremony. If I'm going to go through this, I'm going to go with the guys that I did it with fifteen years ago.

So I get the whole crew in. I bring thirty-two service members to the ceremony in the East Room at the White House in June 2019, including twelve who were there with me on that night in 2004, plus five Gold Star families, the interpreter, and Mick Ware.

I served with some of the greatest men I've ever met in my entire life. And I truly believe that 99 percent of our military is Medal of Honor capable. Any soldier who is put in a position to bleed in order to save people would do exactly what I did. This is who we are and how we were raised in the Armed Forces.

There are a million reasons why we're divided in this country, but I've never cared what your skin color was, who you worshipped, how you voted, or who you loved. Male, female, if you are willing to get shot at for me and my buddies, I will follow you, and I will lead you anywhere. We're family.

That's what makes us elite American warriors. When I was younger, I thought I needed hate to win, hate and anger at my enemy to sustain myself. Now, as I look back, I recognize that we don't fight out of hate. We fight for love — love of our country, our homeland, our family, and our unit. That's stronger than anything the enemy has.

Chuck Pfeifer

Green Beret, US Army
Special Operations Group
Conflict/Era: Vietnam War
Action Date: August 23, 1968
Silver Star

I'm lying in a hospital bed in Da Nang. It's July of 1968 and I'm suffering from malaria, my body wracked with extreme muscle aches and crippled by a fever that, at times, reaches as high as 106 degrees. I'm shoved into ice-water baths that leave me shivering for at least an hour afterward.

I'm a Green Beret and member of the highly classified special operations unit MACV-SOG, the Military Assistance Command, Vietnam — Studies and Observations Group. SOG is so top secret, its existence is denied by the government. We work in small teams to carry out classified ops behind enemy lines. You name it, SOG does it. We don't wear dog tags or identifiable uniforms. We carry non-standard-issue weapons.

I've conducted dangerous reconnaissance missions in the jungle, trying to locate missing teams. Some have been successful; other times I've found only the bones of missing soldiers. I've headed up a group of Chinese "nungs"—mercenaries hired by the CIA to run down the North Vietnamese. I've commanded a hundred or so nungs recruited from nearby territories—they're loyal, serious fighters—and I've planted explosives in Laos and managed not to get myself blown up.

Which is why I feel invincible. And now, after nine months of action in-country, I've been taken down by a goddamn mosquito bite.

I shouldn't be stuck in this bed. I shouldn't be here in Vietnam, really, given my background, but my life has taken all sorts of crazy twists and turns.

I grew up a privileged kid on New York City's Park Avenue. Lived in a penthouse. Got kicked out of Dartmouth after I put an axe through a dorm room door because... let's just say I suffer from too much testosterone. I could go back to college if I gave Uncle Sam two years of my life, so I joined the Army, played football as a running back for the US Army team at Fort Dix, which got the attention of West Point.

I went to West Point Prep School first and entered West Point in 1961 with dreams of becoming a football star. A knee injury killed that dream. As for academics—I admit, I was a shitty cadet. Graduated at the bottom of my class. I chose infantry. Not that I had a lot of options.

I *did* graduate number one from my Ranger class, then I went on to complete Airborne and Pathfinder schools. I've gone to Underwater Swim School in Italy, worked in Germany with the Navy SEALs, trained with Special Forces guys from all over the world—I lived it all and saw it all, loved the action. Then the Army told me I was going to Vietnam.

I wasn't excited by the news, not like some of the guys I know who couldn't wait to get over there and fight. But when I did fight, I realized how much I loved it. Combat is a lot like football—the same adrenaline. Kill or be killed. I see death and destruction all around me, but I don't think about it. I'm not worried about getting killed. I love the action and would gladly sign up for a second tour.

For now, I'm out of action—and out of *the* action—bedridden and unable to sleep, shivering and sweating and suffering from delusions day after day after day.

The next month, August, I leave the hospital forty pounds lighter. I take a desk job at Forward Operating Base (FOB) 4, a top secret special operations base. Across the road is a beach with white sand and waves that would attract surfers.

On August 23, 1968, at two o'clock in the morning, I'm sleeping in my tin hooch— my living quarters—and wake up to gunfire from AK-47s and CAR-14s and M79s and the sounds of mortars whistling through the air and exploding around our base. I immediately grab the Browning nine-millimeter I keep underneath my pillow. I look out the screen window right above my bed and come face-to-face with a sapper.

"Sapper," from the French word for "sap," originally referred to French soldiers—France colonized Vietnam in the late nineteenth century and maintained control until World War II—who dug narrow trenches, or "saps," toward an enemy structure to move men and artillery. Sapper is our name for the enemy's elite Army assault force. North Vietnamese sappers are fearsome and deadly and have no problem sacrificing themselves. They often wear white headbands lettered "We came here to die." The words are written in blood.

The sapper outside my hooch is wearing breechcloth and a gray bandanna. As I shoot him in the head, a door creaks open, and then I hear the unmistakable clump-clump sound of a hand grenade rolling across the floor. I manage to pull my mattress over me before the grenade goes off.

The explosion knocks me out of my living quarters, into the sand. I'm naked and barely conscious and can see sappers running by me and carrying wicker baskets full of grenades as they head to the other side of our base. It's pure chaos. Later, I'll find out they attacked tonight because the new moon cycle would guarantee them total darkness.

It takes me a moment to come to my senses. My ears are ringing from the grenade explosion, and while I've never been more terrified, the adrenaline is giving me the "I don't give a fuck" factor. I run back inside my hooch and put on a pair of shorts. I grab my flak jacket and my CAR-15 submachine gun, link up with a buddy, and go to war.

I run barefoot through intense crossfire. Buildings are demolished. Some are in flames, and everywhere I look I see our dead and wounded. I find three sappers hiding inside a toilet and blow them to pieces. It's a horrible sight — one I know I will never forget.

I assemble a small fire team and make my way slowly to the outside of the camp while rescuing as many soldiers as I can. Then I call in air strikes. My favorite airplane is the A-1E, not the jets. Jets are good for maybe fifteen or twenty minutes before they need to leave, but the A-1E's, which are prop-driven airplanes, can stick around longer, carry more ordnance, and have six 50-caliber machine guns that can shoot the shit out of the enemy.

A North Vietnamese commando team is dug in at the entrance of the camp. I fight my way there, but I can't get to them. I have only a couple of grenades with me. One of the SF guys who has linked up with me has more.

Using my West Point football moves, I start lobbing grenades at the enemy. A couple of bad guys try to run away, and we shoot them dead. I end up killing them all. The hand grenades I used — it's the defining factor in helping save the camp.

At 6:00 a.m., I'm sitting in FOB 4's bar. I have a shot of Johnnie Walker. The NV commandos did their research, knew exactly when to strike. There were roughly a hundred or so Special Forces at our top-secret base, some top generals, and some Secret Service people. Seventeen Green Berets and forty nuns have been killed. I've lost count of how many are wounded.

Five weeks later, I come down with hepatitis and a recurrence of malaria. The pain is incredible, just awful. The single worst experience of my life.

AMERICAN HEROES

This is it, lights-out. It's over.
I'm sure of it.

Somehow, I'm alive. I return stateside.

A friend of mine named Ted has a farm that's not too far away. On the weekends, I ride my motorcycle to his place and do some manual labor to help work some things out physically—and mentally. Manual labor is one of God's great therapies.

On Christmas Eve, my dad and I are flipping through a book of Vietnam photographs. All these emotions kind of well out of me. And I break down into these deep, deep sobs.

Here I'm home for Christmas and a lot of kids aren't home, that didn't make it, or are really fucked up.

The killing I did over there... that's not what's getting to me right now. It's all the bodies of my fellow soldiers, seeing them stacked up like cordwood.

I think back to when I was at the hospital suffering from my first bout with malaria. I met a young guy—a lieutenant—whose eyes had been destroyed by mortar fire. One day, when I was feeling better, I pushed his wheelchair around the grounds while a band played "God Rest Ye Merry, Gentlemen." I almost broke down, but I quickly shut my feelings off, which made me wonder if I was still human.

Then there was the day when I wandered into an airport hangar. I shut the door, and the area turned pitch-black. I struck a match and saw the bodies of three hundred Marines, bagged and tagged and ready to be transported home. There are more memories—finding the bones of missing soldiers, pilots from

SOG-related missions who are still MIA—and they keep coming after me.

I need help. I don't want help. Right now, I'm content with booze and cocaine and my job in advertising and my new wife. My Vietnam pay went straight to my stockbroker brother, who has turned it into a tidy sum of money.

I fell in love with the war. The adrenaline rush, the outlaw feel I got on my top-secret jungle missions, everything. I buy a big Harley and at night, while the city sleeps, I drive up Sixth Avenue going eighty miles an hour. It gets my blood pumping, gives me the rush I so desperately crave every single day.

I become the Winston cigarette man; my face is plastered on billboards all over Times Square. My marriage dissolves in 1980, and I spend the rest of the decade drinking and having too much sex and cocaine. One night I play Russian roulette with a good friend. I'm furious when his wife breaks up our party.

I become a TV and film producer and get small acting roles in Oliver Stone's *Born on the Fourth of July* and *Wall Street*. I do a music video for the singer Prince. I'm also doing a little writing. I'm a regular contributor for *Interview* magazine.

I'm going to do a piece about Vietnam for *Esquire* magazine. Two decades later, I'm back in Vietnam, along with my girlfriend. The country is completely different now, but all I can see is the me—the young soldier—who was here during the war. Everything scares me. I go through waves of terror. I manage my anxiety attacks with booze and Xanax.

My tour guide is a Vietnamese woman who is high up in the

chain of command at the press office. She has arranged for me to meet Võ Ngnyên Giap, a self-taught soldier who went on to become the general for the People's Army of Vietnam. During the war, he had a price on my head—on the heads of all the SOG people. He tried to kill us many, many times.

And now here I am, face-to-face with my onetime nemesis.

Giap is a very small man. Intellectual and charismatic. I tower over him. I'm soaked with sweat. I salute him and say, "I am a graduate of West Point."

"I never studied in military college," he replies. "I acquired knowledge through experience... knowledge I got in exchange for bloodshed and sacrifice."

He doesn't seem interested in discussing the war—discussing anything really. He seems more interested in my girlfriend.

As I'm about to leave, he says, "You know, I don't like you very much."

"I don't like you much, either."

I can't leave fast enough.

I own a bar on the Upper East Side called the 1629. It attracts a lot of tough guys—Hell's Angels, mob guys, and Irish gangsters. Some cops, too, and Special Forces guys when they're in town. Wilt Chamberlain is a regular.

I leave the bar one night and see a rumpled, down-on-his-luck vet. I feel sorry for him and give him some money. For whatever reason, this moment is a turning point.

I've got to change my life.

I find a psychiatrist who treats war vets—especially those

who, like me, are suffering from PTSD. I didn't even know it was a thing until sometime during the late eighties, maybe even the early nineties. I spend the next ten years on his couch, and although it costs me a lot of money, it saves my life, helps get me off drugs and booze.

Soon after I came home from the war, I was awarded the Silver Star, Bronze Star, Air Medal, and Purple Heart. To tell you the truth, I was hoping to get a Distinguished Service Cross.

I no longer give a shit. Medals are medals. Here I am, alive and sober and talking about the war. My life is good.

Paul Zurkowski

104th Expeditionary Fighter Squadron Commander
Lieutenant Colonel, US Air Force
Maryland Air National Guard
Conflict/Era: Operation Enduring Freedom
    (Afghanistan)
Action Date: June 28, 2012
Distinguished Flying Cross with Valor

The troops on the ground need close air support. Fast.

Ninety coalition soldiers are pinned down on a mountainous ridgeline south of an east-west-running river. Taliban forces are positioned on high terrain across the river from the friendlies.

The area is too hot for helicopters to extract our soldiers from a helicopter landing zone two thousand meters south in elevated terrain. Three aircraft are orbiting overhead in the stack from twelve thousand to eighteen thousand feet—a B-1 bomber, an AC-130 gunship for fire support, and an MC-12 Liberty for reconnaissance support.

My wingman is Major Christopher Cisneros—call sign "Metro." He and I have been briefed on this particular mission—the troops we would be supporting, the threats they would most likely encounter, the weather, everything.

The Hindu Kush mountain range is nearly five hundred miles long and as much as 150 miles wide, stretching from central and eastern Afghanistan into northwestern Pakistan and far southeastern Tajikistan. We have reviewed the operation's geographical areas of interest on five GRGs—Grid Reference Graphics—and know it well, having fought there the day before.

I'm stationed one hundred miles away at Bagram Air Base. My plane, the A10 Warthog, can do six miles a minute, so the target area is ten to fifteen minutes away.

At ten o'clock, we take off with a "full bag" of gas—eleven thousand pounds of fuel. It is hard to get into trouble flying this user-friendly, highly maneuverable airplane loaded with systems including a targeting pod that is employed to maintain situational awareness by tracking the friendlies and enemies on the ground.

Our takeoff weight is nearly forty-five thousand pounds, four thousand of which are weapons and ordnance hung on six of the eleven hard loading points. The Warthog's Avenger gatling gun system extends from the nose of the airplane and holds 1,150 30 mm high-explosive incendiary rounds with a fire rate of sixty-nine rounds a second, or forty-two hundred rounds a minute. For this mission, we are also carrying four five-hundred-pound GPS-guided bombs called Joint Direct Attack Munition, or JDAM.

* * *

To reach our target area in the Kunar Valley, Metro and I need to clear mountains topping out at twenty-one thousand feet.

As we fly low-level at eighteen thousand feet, I think about the local people who live in huts and grow and hunt their own food. No UPS deliveryman coming with packages. No electricity. These people live completely off the grid.

The weather is clear when Metro and I arrive in our Warthogs to provide armed overwatch. Gaining orientation in the mountainous target area requires careful assessment. A large tributary of the Kunar River splits into a smaller tributary system flowing toward the Pakistan border, where the terrain is about eight thousand feet.

Two Joint Terminal Tactical Controllers (JTAC)—call sign "Mayhem"—are on the ground. In the process of trying to get all their guys up this high terrain south of the river, they rattled a bee's nest and now they are encountering heavy resistance from enemy guns and personnel.

Looking through the targeting pod, I can occasionally make out movements in the rocks, but mainly I'm coordinating with a JTAC and briefing my wingman on which Y-shaped intersections in the tributary system correspond to which GRG.

We've been flying in the target area for almost an hour when a system of bad weather to the north with bases around fifteen thousand feet slowly moves south and the ceiling, or base of the clouds, decreases. The B-1 is the first one to check out and depart the area, followed by the AC-130.

I tell Metro to depart the target area to rendezvous with the orbiting KC-135 tanker to refuel. The ceiling decreases to five thousand feet, and the MC-12 also checks out of the stack, unable to support the mission due to the weather. I'm now operating a single ship.

Metro radios me. "I'm on my way back," he says. "The tanker is still in the southern area of operations. It is going to complete refueling another set of fighters and then the tanker will come north to refuel me."

"Metro, you've got to go to the tanker or we're both going to run out of gas." Metro heads south and meets the tanker in clear air and coordinates an ad hoc tanker track south of the rainstorm.

Now it is starting to rain in the target area, hard and heavy, to the point where I can no longer see the high terrain to the southeast. It is also starting to get dark. The enemy thinks our air support has left, so they decide to launch a full-on ground assault.

Our Special Forces guys from the UK's Special Boat Service (SBS) start taking heavy fire from the west and the north. They request a show of force, which is a low pass to show the presence of airborne fighter support.

I drop down into the valley and fly down the river between the enemy and the friendlies. This will be the lowest ceiling I have ever operated in, but I have been trained to operate down to one hundred feet AGL and maintain situational awareness by referencing the terrain. Fortunately, I have been in the target area for over an hour and have some familiarity with the terrain. I have probably less than a mile of visibility.

I hook left and fly down the river tributary to the south. There is another Y-shaped split in the river to the west. This show of force ends up breaking the enemy fire coming from the ridgeline to the west.

Our guys are still receiving significant fire from the north and ask me to put some ordnance down there on the high ground.

The rules of engagement say I cannot endanger any civilians or — more importantly — hit any structures because we don't know if noncombatants are inside.

"Stand by," I say as I plot the coordinates. "Let me put down a couple of rockets just to confirm we have the correct location."

Visibility remains low, but my comfort level rises when I get what is called a "nine-line brief" and the tactical restrictions include "all friendlies are south of the river and to keep all fire north of the river."

I roll in for the pass and get clearance to fire. I fly close to the target and fire only one rocket. I am also taking on a barrage of small-arms fire, and I see the tracers go past the canopy. The rocket confirms the target location and Mayhem requests 30 mm on that location.

As I get the next "nine-line," I check my fuel and see I am not going to be able to make it to the tanker, as Mayhem needs coverage.

Our minimum fuel is twelve hundred pounds and I will need a thousand pounds to make the 100-mile flight back to Bagram. Ideally, we land with at least two thousand pounds of gas, so I reset my minimum fuel "bingo" to about twenty-two hundred pounds.

Now I'm going to execute my first gun attack.

As I begin my second pass, Metro radios that he is refueling with the tanker. When I roll in, I can still see the terrain through the canopy, but I cannot see the target through the thick front windscreen capable of taking a hit up to a 37 mm round. This is due to the encroaching darkness and heavy rain.

The Warthog's targeting system has what is called "Mark Zulu." In the heads-up display (HUD), I can pull up a mark of the impact point from my last attack—in this case, where the rocket hit the ground. I can offset from that Mark Zulu cue using my gun reticle in the HUD.

I put down about two hundred rounds on the ridgeline. It gives the guys on the ground the opportunity to move farther up the terrain.

As an attack pilot, I am looking for battle damage assessment, or BDA. The JTAC is not concerned with BDA, as this is more of a fire and maneuver event. They are looking for rounds to get the enemy's heads down, so they can move up the ridgeline.

On my fourth pass, Metro tells me he is eight minutes away from the target area. Later, the guys on the ground will tell me it is like I just appeared out of the clouds with my strafe passes, the visibility is so low.

I pull off my fifth and final pass. I am out of 30 mm and below my twenty-two-hundred-pound Bingo fuel—only seventeen hundred pounds left. I am now out of emergency fuel. I inform the JTAC, as I climb into the cloud deck to return to

Bagram, that it will be roughly seven minutes until there is another A-10 on station. There will be a break in coverage.

I declare emergency fuel.

The way to "make gas" is to climb the aircraft at L/D max. The maximum ratio of Lift (L) over drag (D) delivers optimal climb—and descent—speeds. I pull up into the bad weather and keep climbing through it. I need to get as high as I can to save gas and clear the terrain. Given the Warthog's weight, I climb until I reach twenty-eight thousand feet.

Now I can cruise at altitude and plan an idle descent back to base trading, potential energy for kinetic energy. But there's a problem. The weather extends to the west over the high terrain. I need to stay higher longer to descend in visual conditions—and I burned more fuel at altitude than planned. Fortunately, Bagram is in the clear. I descend with Bagram in sight and land with eight hundred pounds of fuel.

While I'm airborne en route to Bagram, the guys on the ground need coverage. I try to get as much info as I can to Metro, who will be taking over the target area. He tells me he is about four minutes away and I relay the information to Mayhem, our JTACs on the ground.

Metro calls me and says that he does not think he can make it back in below the weather due to the high terrain.

"This is what you need to do," I tell Metro. "Go to the south end of the Hindu Kush Mountains. Drop down there and then come up the river and it is going to be the second or third tributary on the right. But you are going to have to come in low-level below the weather."

"Okay," Metro replies. "I can do that."

I work with the Air Support Operations Center (ASOC) to reroll some more A-10s from the Southern Area of Operations to support Metro, then I try to coordinate for a "quick turn of the gun"—get refueled and reload the gun system—at Bagram and then get back up in the air.

"I think you want to come into the building," I am told. "The Wing Commander is waiting for you."

I land safely. The A-10 is covered in gun gas. It is going to need a good washing. I hand the plane off to maintenance and go to see the Wing Commander for my ass-chewing for being on emergency fuel and operating in low weather.

First, I go through a debrief with maintenance and they report they found two bullet holes in my aircraft. I knew I had been shot at, but I did not know I had been hit. I then debrief with Intel and go to see the Wing Commander.

"We have got an issue here," the Wing Commander tells me. "Both JTACs have been injured on the ground. I need you to get word to the pilots in the target area they are cleared to control the fight, if they are FAC"—Forward Air Controller—"qualified." The rules of engagement state you cannot expend ordnance without a FAC or JTAC. The International Security Assistance Force commander does not want airborne FACs controlling the fight because of the potential for civilian casualties and not being able to see the full picture from the air.

It turns out there is a third JTAC on the ground who is right out of school. He has only controlled aircraft during training, and now he is single-man ops running the fight.

With the Warthog's targeting system, we can plot where the friendlies are and mark their locations on our moving map. When Metro returns to the dynamic target area, the enemy is where we had marked the friendly location before he left for the tanker. This confusion is resolved by the JTAC's brief as the fight is now danger close. The enemy is within one hundred meters of the friendlies and 30 mm fires are required immediately.

The third JTAC is also carrying a rocket-propelled smoke round for marking targets. He fires the rocket to mark the enemy location and Metro puts rounds down. Two other Warthogs join the fight and execute coordinated attacks until the enemy is neutralized.

Additional A-10s relieve Metro's three ships and stick around and provide cover for the HH-60 Pave Hawk casualty evacuation helicopters to land and rescue our ground troops. Some soldiers are wounded, but every single man is alive.

The two injured JTACs are flown to the forward operating base. From there, they are transported to Bagram. Gurneys are waiting to take them into the hospital.

I am told they insisted on walking.

The next day, I go see the two JTACs, along with the Vice Wing Commander, the Ops Group Commander, and Metro. We want to make sure they are okay.

One guy who was shot in the chest is still in surgery. The

other was shot in the arm. He looks at me and Metro and asks if we are the two A-10 pilots who stayed when the weather got bad.

"We are," I say.

"Brilliant!" he says. "You saved a lot of lives."

Meeting one of the guys we helped save and hearing him say, "You are the reason I am alive today"—there is no greater satisfaction for an A-10 pilot.

Arthur Rice

First Lieutenant, US Army
Conflict/Era: Vietnam War
Date of Action: April 25, 1969
Distinguished Flying Cross

I'm a fuckoff and wouldn't do things any differently.

I've been at two different colleges in two years. I think I have fifteen credit hours, three of which are half hours, and the US is drafting anywhere between forty thousand and fifty thousand young men per month.

I'm young, naive, full of moxie. I'm not the least bit concerned about getting drafted. If I do, it won't matter. I'm Superman. I'll just do whatever I need to do and then come back home, pick up where the party left off.

I get drafted. I decide I'm going to outsmart everyone, so at the last minute I enlist in the Army as a clerk typist. I take a train out of New York City to Fort Jackson. On the way down to Columbia, South Carolina, the guys with me trash the entire train down to its steel wheels and the frame of the compartments. Got to love New Jersey guys.

Fort Jackson is kind of interesting. Guys are dying from spinal meningitis, and no one can figure out why. There's so much stuff going on in 1966 this news doesn't even make the papers.

I'm still signed up for this clerk typist thing, and I'm given some tests. I sign up for Officer Candidate School (OCS). I go through basic, then Advanced Individual Training in Alabama. From there, it's off to OTC at Fort Benning, Georgia.

During OCS, we're out in the boonies about half the time. They call it "being out in the field." There, they come around and say, "The Finance Corps is coming. Who wants to go listen to their pitch?"

I go because I'll be put on a bus and then delivered to an air-conditioned room where I can drink Cokes and smoke cigarettes. Anytime there's a chance to listen to a pitch — Finance, Chemical, Transportation, anyone — I always put my hand up.

Then along comes flight school. The test they give me is interesting. It doesn't measure intellectual capability. It's more interested in spatial orientation — my ability to figure out whether or not I'm upside down. I'm guessing that's important when flying a helicopter.

Something in me changes after OCS. I'm as gung-ho as you can get. Ready, willing, and able. I'm given a choice between Ranger training and flight school. I choose flight school. There are helicopters, and guns and shit. I'll go over to Vietnam and fly helicopters.

I graduate flight school in November 1968. I arrive in Vietnam on December 23, 1968.

Sergeant Major Bolin of the 3rd Squadron, 17th Air Cav, picks me up at Bien Hoa, takes one look at me, and, with a smooth Southern drawl, says, "*Loootenant Rice,* you can't see the colonel lookin' like that." I'm told to get a haircut.

The 3/17th is comprised of one ground troop, Delta, and three Air Cav troops, Alpha, Bravo, and Charlie. The colonel makes me a platoon leader with Delta Troop, acting as air mobile infantry. I'm the only infantry officer in the squadron. None of my guys have infantry training. Everyone else is an armor officer, but Delta Troop is acting as air mobile infantry as a ready reaction force for these long-range recon patrols.

The LLRP units — Long Range Surveillance Detachments — we're working with always have teams out in the field. They're not supposed to make contact, and they always make contact.

"When you go into a landing zone," I ask them, "how do you secure it?"

"We don't."

"If you don't secure it, then how are you going to get out?"

They don't know because they've never been trained. I work with them for a couple of months. We don't make contact once. The most interesting thing we do is blow up a hospital bunker complex in an area called War Zone D, located in 3 Corp.

Then I become the XO. I don't want to be an XO for Delta Troop or anyplace else because I'm *really* gung-ho now, chomping at the bit to get into the fight. I tell the sergeant major, who talks to the colonel. He sends me down to Alpha Troop to start flying our hunter-killer teams—low-flying OH-6 Cayuses drawing fire away from AH-1G Cobra attack helicopters.

The OH-6, or "Loach," short for "light observation helicopter," is shaped like an Easter egg. The minigun has up-and-down elevation but nothing side-to-side. The ammo box is behind the two front seats, which are armored. The floor is not.

Charlie Troop flies with the same configuration, plus a third guy in the back that has access to an M-60 hanging from a damn bungee cord. I start flying with them. One time, the wise guy in the right seat starts screwing around, flying in such a way to induce motion sickness so I'll throw up. Motion sickness is one of the few weaknesses I don't have—and I'm more concerned about the guy in the back with the M-60 on the bungee cord. I'm worried he's going to accidentally blow my head off.

In early March, we do missions with two teams on station—two Cobras and two Loaches. We're at the north end of the Delta, southwest of Saigon.

Six weeks later, in April, we receive intelligence warning that a pretty good-sized Vietcong force and possibly some North Vietnamese Army regulars are in the area, making use of a network of underground tunnels. The Air Force comes in and completes a bomb strike. Instead of using high explosives, the bombs are loaded with CS gas.

Because CS gas is heavier than air, it stays low to the ground. The belief is that the gas will ultimately find its way into the tunnels through airholes. Our job is to fly into the gas, try to detect movement, and use smoke grenades to mark the location with colored smoke so the Cobras can fly in and drop ordnance.

There are several flaws with this tactic, the first one being that the enemy will know our location long before we know theirs. Also, the OH-6s we're flying have no doors, so we'll be wearing gas masks while trying to detect enemy movement in elephant grass that's anywhere from six to eight feet high.

When they start crawling out from their holes, they're madder than hell and a little crazed, because if you inhale enough CS gas, it makes you a little nutty.

We start taking fire. Rounds come up through the floor, and instrument panels start exploding. I've been hit in my right leg, and all hell is breaking loose.

I'm in the left seat with all the maps, and I have an M-16 with a collapsible stock. I lean out of the helicopter and start shooting. We're taking a lot of hits, but so far, nothing critical. The pilot — the warrant officer — pulls in the collective to haul ass and get us out of here. I keep shooting and even up the score a little bit.

I'm losing a lot of blood from my femoral artery, and I'm doing everything I can to staunch the bleeding.

Eighteen minutes later, we land at the 3rd Field Hospital in Saigon. They have a line of stretchers — they're doing box office business here — and when I finally go in, the doctors do an arterial graft. They take out a big piece of the femoral artery in my left leg and then try to graft it to the wounded area of my right.

The graft doesn't hold. I keep springing leaks over the next two weeks. Worse, I'm starting to get gangrene in the front half of my right foot. They do what's called a Syme amputation, then they're forced to amputate my leg about three-and-a-half inches below my knee.

Two days later, I'm in a hospital ward in Japan. It's one of the most terrifying experiences I've ever had. All the guys in here have serious gunshot wounds. Many of them are in casts—some in full-body casts. The people here are in real pain, and there's a lot of screaming when they have to cut "windows" in the casts to change wound dressings.

Two weeks later, I'm stabilized. I'm put on a plane and flown to Walter Reed.

All I can think about is my right leg. That it's gone.

How am I going to go to the beach? I don't want people staring at me. I don't want them to feel sorry for me. How am I going to deal with this? What are women going to think?

These are the thoughts occupying the mind of a twenty-two-year-old.

I feel lost. Like I'm floating through life. I'm sure this sort of stuff happens to a lot of guys who've gone through combat. It must have some effect, right? I know it affected the warrant who flew me to the hospital. He got nicked in the thumb—not the sort of wound that would take him out of the action—but he packed it in. Said he's never going to fly again.

Ultimately, I decide to go back to college, only this time I return with a vengeance. I have some anger issues, and I'm actively trying to drink myself into an early grave. I'm becoming a stone-cold alcoholic.

I finally get to Alcoholics Anonymous. I've been sober a long time now—thirty-five years—and I don't have any complaints or regrets about my military service. I wouldn't do things any differently.

Michelle Saunders

Sergeant, US Army
Conflict/Era: War on Terrorism (Afghanistan)
Date of Action: May 1, 2004
Purple Heart

I'm one of the few females in a male-dominated job, by choice. I challenge myself. I prove myself physically, enough so to challenge the guys around me to be better. I'm focused on being a great soldier, as well as an all-Army athlete. I love, absolutely love, the culture and the environment.

My family is patriotic, but it's also broken. My brother was killed in a drunk-driving accident during my senior year in high school, and it sent us all into a tailspin. His death made me reconsider my plans to attend college. I was being heavily scouted to play D-I softball, but I needed to get away from me, the family who raised me, and find a new and different one.

I came into the Army as a 62-Echo, which is a heavy equipment operator. I've switched over to a 21-Echo, which is basically engineering. I would love to go into Special Forces, but women aren't allowed.

In 2001, I'm stationed at Fort Campbell, Kentucky. I finish a ten-mile run and head back to my place in Oak Grove, three and a half miles from the base. I turn on the TV and eat my cereal. The screen flashes with breaking news in New York City. A pilot's flown his plane into the World Trade Center.

This is crazy. How did this asshole hit a tower with a plane?

A second plane hits.

I drop my spoon into my cereal bowl and say, "Holy shit, that was intentional."

My phone starts to blow up. I'm ordered to grab my gear and get onto base, which is going into lockdown. It takes me almost six hours to move through the security checkpoint.

Everyone inside the base is already exhausted from the chaos. All the comms are jammed with family members counting on us for information. They're crying and confused because they don't understand what's happening, but we don't know much more than they do. All eyes are glued to the TVs running nonstop while soldiers do a full inventory of their gear.

I prepare to say goodbye to my family and friends.

My battalion is told we're going to Kandahar, Afghanistan. I'm hyped up and ready to go but equally scared. I know it's just a matter of time until I get my orders to go.

Right as we're about to deploy, a bunch of us get different orders. We're going to Hawaii for further training to prepare for

Logistical Support Area Anaconda, Iraq, and improve force protection.

Two years later, in 2004, I'm on a plane flying to Kuwait. I'm excited but nervous. It's the big game on the biggest stage.

You've trained for this, I keep telling myself during the flight.

At the base camp in Kuwait, I settle in, getting acclimated to the desert climate while also preparing myself mentally and physically for the long road to Shitsville. Balad Air Base is forty miles outside Baghdad. Our convoy is one giant, moving, unarmored target. We don't have scouts or MPs going ahead of us to check things out, and we have zero air support. I feel vulnerable, defenseless. My ass starts to pucker up and my head is on a swivel, looking everywhere for unseen threats—roadside bombs set to explode on contact or shooters' positions to take potshots at us.

It's the longest ride of my life.

Our mission is to strengthen the defenses of the base camp: build up the trenches and the Hesco barriers serving as berms; safeguard the living quarters from incoming attack. We'll also be setting up the security checkpoints from Kuwait and on the main supply routes through Taji, the Sunni Triangle, all the way up to the Green Zone in central Baghdad.

The HESCO barriers come in pallets. We put them together like Lego sets, then fill them with sand. It's the suckiest job on the planet, but the end result is worth it.

Engineering units are self-sufficient. We have our own fuelers. We pull our own security, run our own convoys, and we're literally pulling metal from scrapyards and welding the pieces onto our unarmored vehicles.

AMERICAN HEROES

There's always excitement on the road. Potshots from snipers and a couple of IED scares. The enemy knows the exact grid coordinates of every part of the base. I wonder if it's because the US pays local nationals to come in and do daily labor.

Intellectually, I know that not every Iraqi citizen is an enemy. But when you're a soldier in-country, you think everyone's trying to kill you. Trying to decipher who's who every second of every day is mentally exhausting. Some soldiers are locked in a hypervigilant state.

It's tough to break out of what can become a paranoid mindset.

Anti-American rage is fierce, especially in and around Fallujah, Iraq. At the end of March, four Americans working for Blackwater Security Consulting are ambushed and killed. A joyful mob drags their burned bodies through the streets and hangs their corpses from a bridge. Fifteen miles away, a roadside bomb kills five American soldiers.

One night, when we're getting ready to roll out, a soda truck shows up at our main gate. The driver is well known to us and makes deliveries to the base once a month. He always waves and says hello to everyone.

Tonight, the entire base camp is rocked. Sirens start going off. Everyone jumps into full battle rattle because we think we're under attack. We are.

Lo and behold, the soda truck driver was a suicide bomber.

We never find out what motivated him to turn against us. He could have been threatened, something like, "Hey, we have your family and we're going to kill them unless you do what we tell you to do." We'll never know.

I don't get any sleep that night. None of us does. The next morning, as we're getting ready to roll out to Kuwait, a two-day trip, I turn to one of the convoy leaders and say, "Man, I just don't have a good feeling about this. It doesn't feel right."

"It's going to be fine."

Our first checkpoint is the Abu Ghraib prison. This scandalous place has made international news—that Saddam Hussein used the place as a torture chamber and also that US military police abused Iraqi prisoners here.

We get into the prison and stage our trucks—all thirty-six of them. Immediately, we're pummeled with incoming fire. They bombard us with rockets, RPGs, all kinds of shit.

We're all going to die.

Everything goes slow-motion. It's the most bizarre feeling. And all those clichés about facing death—your whole life, from your earliest childhood memory through all the moments up until now, flashing in front of your eyes—all those clichés are absolutely, 100 percent true.

Things finally calm down. Somehow, we make it through the night without taking on a single casualty, but we're sleep-deprived and scared.

The prison is located about twenty miles west of Baghdad in the middle of the desert, where daytime temperatures can rise to 140 degrees and nighttime temperatures drop to the seventies. This morning, fog reduces our visibility to near zero.

Our main supply routes, we're told, have been blown out. As we start to map out alternate routes to Kuwait, we receive word we'll have air support. I'm hugely relieved—until our air security cancels due to lack of visibility.

We're going to have to wait it out and hold on for another day. That's my assumption, but the commander has another idea.

"We're going to continue on with the mission," he tells us.

Everyone is pissed off. But we have our orders, and we get on the road in the thick fog. As soon as we leave the gate, one of our vehicles gets T-boned by a civilian dump truck. The driver couldn't see us. Three of our guys are taken out. One guy's leg is snapped in half. He needs to be medevaced.

It's a shitty mess.

The towns are super small, maybe five hundred to one thousand yards long. We roll through one—our guys ready to get on top of our trucks to "pull the pole" from low-hanging telephone lines.

And the locals are *pissed*. We slowly escort, on foot, in front of the trucks with our weapons, surrounded by hundreds of people at any given time. We're all vulnerable. If they want to take me out, they totally can.

In the desert, we're driving through the middle of nowhere until late afternoon, when we reach our third checkpoint, a little town right outside Basra, Iraq. A little girl comes out waving her arms. Her movements and the expression on her face are telling me, *Don't come here. Don't come in.*

I'm in the third vehicle, riding security in the "gunship," a Humvee with an M249 light machine gun mounted to the turret.

We come to a stop. It's time to do a rotation. Medina, one of our mechanic sergeants, jumps into the gunship. I head to a Mack truck hauling a grader and assume the tank commander position.

We get to Al-Amarah. That's when all hell breaks loose. We start taking on crazy heavy fire, rocket-propelled grenades — the whole nine yards.

How can this be happening? There are checkpoints coming in and checkpoints going out, and they're supposed to be manned by local nationals who are on our side. Clearly, they've been overrun by the enemy — and there's no way they're going to be able to help us.

The enemy was lying in wait. Now, it's a full-on ambush. Complete chaos.

We're screwed.

Medina, the mechanic sergeant who took my place in the turret, takes a round through the side of his vest.

That would have been me.

To see one of your soldiers wounded, let alone killed — there's no training for that.

One of the other mechanics, Specialist Ojeda, takes a round through the helmet; I rush to him and Medina. The area is quickly turning into a kill zone, and we need to get out of here — *fast.*

Without his gear, Medina is a good 250 pounds. I pick him up, and when I attempt to throw him over my shoulder, I do something to my back — something bad. I'm not feeling any pain yet because I'm running on adrenaline.

This eerie stillness spreads through me. I know I'm going to die. Today, I'm going to die, and I'm okay with it.

We get Medina and Ojeda into the truck. It's got a goddamn grader on the back—we're talking five or six tons. How fast can we get out of here? And then we're also going to have to somehow maneuver through the HESCO barriers at the checkpoint.

It's slow moving and we're still getting shot at and we've got all these big trucks. Some of them are damaged, and we need to blow them up because we don't want the enemy to take our intel, and the rest of the trucks are moving too slow, and all our comms are crushed, and I'm sure I'm going to die.

Then I remember I have a satellite phone with me. I bought it so I could talk to my family at any opportunity instead of waiting in line at the internet station on base, where we have less than a minute to talk—and that's provided the call doesn't drop. At this point in time, the internet isn't like it is today.

I use the SAT phone to call our base camp. The people there send out some sort of Mayday. A nearby British element comes to our rescue and saves our ass. We end up going to a British base camp.

The ambush has left us with eleven casualties. Two were KIA, and the rest were injured from multiple RPG shrapnel wounds, all kinds of things. I'm medevaced out with the other wounded soldiers. It's the last thing I remember.

I wake up at our Army base in Landstuhl, Germany.

I've blown out my back, two ruptured discs and nerve damage. I'm in terrible pain, and all I want to know is if everyone is okay. No one here knows anything.

The doctors keep pumping me full of morphine. I keep drifting in and out of consciousness. They're trying to figure out what to do with me. "Can we fix her here and then send her back to base camp?" I hear one ask.

Another doctor says, "She's going to need surgery."

No one calls me. I have zero communication with anyone downrange.

Eleven days later, I'm flown to Walter Reed. When I wake up after my surgery, my parents are there.

"Am I alive?" I ask my mother. "Am I dead?"

"You're alive," she says, dumbfounded.

The person across the hall is missing half of her skull. Doctors put part of it in her stomach so they can keep her brain tissue alive. A guy three doors down lost three limbs. Seeing these massive injuries makes me feel guilty about even being here — being alive. Sure, I'm in a wheelchair and have a few scars, but my head is intact, and I have all my limbs.

My mind, though, is *really* screwed up. I can't grasp reality. Have no idea if I'm alive or dead. I can't stop thinking about the people downrange, wondering if they're okay, and I keep having nightmares about that little girl from the Iraq town who came out and tried to wave us away.

This goes on for months.

My personal belongings are back in Hawaii. The Army tells me they've packed everything into a footlocker that is now in storage.

"What do you want us to do with your stuff?" the guy on the phone asks.

"Can you send it to my parents' house?" Then I ask him about my unit, how they're doing. He doesn't know anything.

The doctors put me on all kinds of medication, including an antipsychotic given to schizophrenics. I can't stand it. Finally, I tell my mom, "I'm not taking this shit." I'll take Motrin unless the pain is excruciating, but I quit everything else and go cold turkey because I want to understand what's happening to me.

I try to call Medina's mother and explain what happened to her son, that I was the one who was actually sitting in his seat before we switched out. The guilt I'm feeling is unbearable. I know I'm going to be carrying it with me for a long, long time.

I spend twenty-two months at the hospital. Near the end of my stay, former US Marine Corps lieutenant colonel Oliver North comes to Walter Reed. The staff hands out gift cards to Ruth's Chris Steak House as North goes down the line shaking hands.

When he shakes mine, he says, "Is your husband wounded?"

His words are a punch to the gut. *Did he really just say that to me?*

"Actually," I say, "I've been a patient here for twenty-two months. But thanks."

"I'm so sorry."

I've had multiple back surgeries. I have PTSD, limited mobility, and all kinds of fricking hardware in my body. The military decides my career is officially over.

It's the hardest breakup I've ever experienced.

I'm homeless. Not living-under-a-bridge homeless but couch-surfing and struggling with back pain and PTSD, and adding insult to injury—literally—the Army has decided not to give me a medical retirement. Instead, they're thinking about giving me severance. I'm completely devastated that the very institution I was willing to die for didn't think I was wounded enough to retire.

I don't have any money, and I'm too embarrassed to ask for help, which is why I've been homeless these past four and a half months. I don't know what my value is anymore and have completely lost my sense of purpose. My résumé is great if I want to serve in the military, but because I can no longer do that, I don't know how to translate my skills and experience to the civilian workforce.

I'm invited to attend a town hall meeting at Walter Reed. A meeting where all the high-ranking brass within the medical ranks discussed how to best support our wounded service members post military career. I decide to stand up and speak.

"I've spent the better part of my adult career in the military. I was a top-notch soldier. I had a career-ending injury, and no one helped me transition out of service. And now here I am, four and a half months out of the hospital, and I'm homeless and I don't know what to do. Here's where you're missing the mark."

I rip into the committee members. I'm diplomatic in my choice of words, but I don't hold back my feelings. When I walk offstage, I'm offered a job by the Department of Labor. The girl who doesn't have a job is given a job as—get ready for this—an employment specialist.

My mission becomes twofold, pushing legislation around the creation of military transition programs, and equally feeling my way through my own transition to whatever is next. What I enjoy most is helping companies identify and hire top talent among military folks, then preparing those soldiers for transition into their first civilian career. It's a gift and a blessing, this work, because it reminds me every day where I came from. The work gives me gratification and a strong purpose.

I'm getting ready to move to a new home. I open the door to my garage and stare at my footlocker. It's been traveling around with me since 2004.

My past is buried in there, but I've never opened it.

A U-Haul is on its way and I've got a bunch of stuff I need to take to the dump.

It's a battle, from that war, that never seems to go away and always leaving me to decide. Am I going to take it with me, or am I going to throw it out?

I stare at the footlocker, wondering what I will find if I lift the lid.

Mark Mitchell

Major, US Army
Conflict/Era: War on Terrorism (Afghanistan)
Date of Action: November 25–26, 2001
Distinguished Service Cross

Six weeks after 9/11, we're putting boots on the ground in Afghanistan.

Everything has been chaotic since the attacks. The US doesn't have any diplomatic relations with Afghanistan or any basing agreements with the surrounding countries. Worse, we have no standing plans for an invasion of Afghanistan under any circumstances, which makes the entire enterprise a perilous situation.

Honestly, on September 10, if you'd have asked any of us about the possibility of conducting operations in Afghanistan, we'd have said, "When hell freezes over!" I never thought I would be traveling to Afghanistan.

Back in late 1993, I was a Special Forces, or SF, team leader.

I had finished the competitive and demanding qualification course (aka, the "Q Course") to become a Green Beret, and my first deployment was to Pakistan. Our Pakistani Special Service Group counterparts took us through the Khyber Pass, up to what's known as the Torkham Gate, which looks directly into Afghanistan. The Russians had suffered a brutal and ignominious defeat there, and I remember thinking, *Wow, I'm glad I'll never go to Afghanistan.* Nobody in the Department of Defense thought that we, the United States, would ever find ourselves operating here, either.

Yet here we are.

In mid-October, nine operatives from the CIA's Team Alpha link up with General Abdul Rashid Dostum, a notorious Uzbek "warlord" with a shady and checkered past. About ten days later, in a remote river valley in Northern Afghanistan, a Special Forces Operational Detachment Alpha (an ODA, or more colloquially, an A-Team) joins them. On the same night, US Army Rangers parachute deep into the country and execute a raid on the complex belonging to Mullah Mohammad Omar, the spiritual and political leader of Afghanistan during the Taliban's rule—and a major supporter of Osama bin Laden. Although combat operations had started with bombing and missile strikes, the raid marked a new phase and for us, it felt like the war for control of Afghanistan had truly begun.

The A-Teams were twelve-man teams comprising Special Forces operators who were highly trained in advanced weapons, combat tactics, demolitions, combat medicine, and working with local forces. From a strategic perspective, their arrival in early October was a lightning-fast and truly incredible response, considering what it takes to move forces overseas, the furious planning and diplomatic negotiations.

Now it's October 26, 2001, and I'm flying to the former Soviet air base in Karshi-Khanabad, Uzbekistan. As a battalion operations officer, I'm expected to set up and run our battalion operations from a tactical operations center, or TOC, safely in Uzbekistan.

It's the middle of the night when we land. We don't have access to any of the buildings on the air base, just tents, and we have to wade through what we call "moon dust," a superfine powder. SF guys are used to austere environments, but I've got to admit this place is pretty miserable. Somehow, though, it's exactly what I expected in a former Soviet republic in Central Asia.

Our technology is also equally austere. The iPhone is several years away, and only the CIA has armed Predator drones, which are still experimental at this point. Our computers are Panasonic Toughbooks, and we'll be using UHF-VHF radios. Even a simple thing like a map is nearly impossible to come by.

Our "body armor" is old flak jackets designed to protect from shrapnel but ineffective against a round from an AK-47. It doesn't matter though, since we won't be wearing flak vests or helmets because our Afghan counterparts in the Northern Alliance don't have them. It's important to share their hardships, to be as much like them as possible.

I find my way to the tent and throw my rucksack and duffel bag on a bunk. I'm about to set out to find my commander when my old company sergeant major walks into the tent and without any fanfare says, "Sir, grab your stuff. You are moving into a different tent. We're in isolation now."

Isolation is the SF term we use for when we're planning an operation. For operational security reasons, teams are put in isolation. They're not allowed to talk to anyone except for a couple of designated officials.

As for what's going on, I don't know much. The Air Force is conducting their bombing campaign, and there're currently two ODAs on the ground—one with the Tajiks, and one with the Uzbeks.

My family has no idea where I am. Because I'm in isolation, I have an opportunity to write letters to my wife and two daughters. If I die, these letters will be the last communication to my family.

My daughters are four and one and a half. How do I explain to them in a meaningful way why I decided to sacrifice my life? Writing these letters and knowing they could contain my final words is truly one of the most difficult and emotional things I've ever done.

When the Soviets retreated from Afghanistan in 1979, various ethnic and tribal factions went to war with each other to control the country. The corruption in the resulting government gave rise to the Taliban in the mid-1990s. A new civil war erupted and by 2001, the Taliban controls almost 90 percent of Afghanistan. The other 10 percent is controlled by the Northern Alliance, the anti-Taliban group composed of various ethnic factions, including Hazaras, Uzbeks, and Turkmans. After the

Taliban recaptured Mazar-e Sharif in 1998, they drove opposition forces into the mountains and cut them off, hoping they would starve to death. These men, who have been fighting to defend their communities, are on their last legs.

The Northern Alliance is an ad hoc coalition, an alliance of necessity—desperation, really—united only in their hatred of the Taliban. It isn't a well-equipped and organized military force by any standard. Their weapons are ancient and worn out; ammunition and supplies are hard to come by. Cut off from the rest of the world, holding out in these remote valleys, their forces are on the verge of collapse. But they are skilled fighters, and with a little support they can be effective on the battlefield.

Though they may speak a different language and have different customs, the Afghan people want what we want: safety for their children and access to education. All other things being equal, the vast majority simply want to be able to live and work in peace and provide for their families. But for decades that peace has proved elusive, and thanks to Taliban support for bin Laden, war is afoot across Afghanistan.

Around November 4, the Northern Alliance, refreshed by arms and ammunition from the CIA and aided by advice and air strikes courtesy of the Special Forces, set off from their mountain strongholds to liberate the city of Mazar-e Sharif from the Taliban. Along the way, General Dostum, a controversial but effective leader of the Uzbek faction of the Northern Alliance, uses the precision air strikes to convince a lot of the Taliban's local partners to stop fighting against him. They won't surrender or abandon their arms but their newly minted neutrality greatly assists our progress toward Mazar-e Sharif. On November 10, 2001, Mazar-e Sharif is the first major city to be liberated from the Taliban.

General Dostum throws a big party for everyone on November 11 — Veterans Day. I'm unable to attend the festivities. As operations officer, I have to work on future plans and keep an eye on the radio back at our headquarters, located inside the nineteenth-century fortress built out of mud and straw bricks known as "Qala-i-Jangi" ("House of War"). About a week later, we move to the Turkish School, a five-story, twentieth-century high school building built by the Turkish government.

On Friday, November 24, General Dostum, the Northern Alliance, and an element from 3rd Battalion head east toward Kunduz, another Taliban-controlled city in northern Afghanistan. Just beyond the city limits, they unexpectedly encounter a convoy of trucks with about six hundred Taliban and Al-Qaeda types, seeking to "surrender." The "negotiations" last most of the day, punctuated by a suicide bomber killing Dostum's intel chief. Despite the ominous treachery, Dostum accepts the surrender of the rest, and they're taken to Qala-i-Jangi.

Dostum and the Northern Alliance forces continue to Kunduz, which everyone expects to be a huge and decisive battle, the Waterloo of northern Afghanistan. The battle never really materializes. After a standoff of several days, with very little fighting and some US air strikes, the Taliban force surrenders.

At QiJ, the prisoners are housed in the southern half of the fortress. In the center is a single-story, reinforced-concrete building we call the "Pink Schoolhouse," built by the Russians during their occupation of Afghanistan.

Before moving to the Turkish schoolhouse, I spent nearly two weeks living inside Qala-i-Jangi. The nineteenth-century fortress is *massive*—over three hundred yards in diameter—and surrounded by a moat and walls that are sixty feet high and thirty feet thick. There are fighting positions on parapets. Most rooms inside the buildings contain vast stores of ammunition and weapons—rifles, machine guns, hand grenades, artillery shells, BM-21 rockets, antitank and antipersonnel mines.

The next day, Saturday, November 25, CIA operatives Johnny "Mike" Spann and David Tyson tell me they're traveling to Qala-i-Jangi to question the prisoners. They're hoping to gather intelligence on the Taliban's key players and maybe even insights into bin Laden's whereabouts.

"Do you guys need any backup?" I ask. "Do you want us to go with you?"

"No, I think we've got this."

Later that morning, we begin to receive reports of gunfire and explosions at Qala-i-Jangi. About the same time, a friend of General Dostum's appeared at our front gate claiming that the prisoners at Qala-i-Jangi have staged an uprising and managed to get their hands on the fortress's stockpile of ammunition and weapons—and now they've seized control of the southern courtyard. But we haven't heard anything from Dave and Mike.

Later, we learn that they were in the process of questioning prisoners in the southern courtyard when they heard an explosion (likely from a smuggled grenade) and were suddenly rushed by prisoners in the courtyard. Within seconds, they were in hand-to-hand combat and gunfire began to crackle around them.

In the chaos and confusion, they were separated from each other. Dave managed to escape to the northern courtyard, which is separated from the southern part by an enormous wall, and encountered a German TV crew member who was there filming. He borrowed the man's satellite phone and called the US embassy in Uzbekistan for support.

Spann's whereabouts are unknown.

Mounting a successful rescue mission will require additional tactical resources. The three ODAs at our headquarters in Mazar-e Sharif, however, have all left with members of the Northern Alliance to liberate the city of Kunduz, and the nearest conventional unit is an infantry battalion from the 10th Mountain Division stationed up at Karshi-Khanabad, which is several hours away by helicopter.

We're isolated and on our own.

Our recently arrived UK SBS counterparts don't have authorization to participate in the hostilities, as their rules of engagement have not been approved. When they learn of what's going on, they tell me, "We're going with you."

We head out in a Toyota Land Cruiser and Land Rovers with general purpose machine guns attached to the roofs. We have no idea what's going on inside the fortress. The gunshots grow louder as we get closer.

When we arrive, after a harrowing, high-speed drive, we manage to find Commander Fakhir, a trusted lieutenant of Dostum. He tries to explain, in Dari, what has transpired. I speak Arabic but not Dari, so I'm at a bit of a disadvantage communicating with him. Nevertheless, I'm able to ascertain that the situation is dire.

The northern part of the courtyard is currently being held by the Northern Alliance. Their numbers are small—roughly one hundred—but they have use of a pair of old Russian T-55 tanks that fire 100 mm shells. If the Taliban seize the fortress and manage to get local forces—the same ones that Dostum convinced to become "neutral"—to rejoin the fight, a lot more Americans will die.

Including myself, I have a dozen men with me—seven SBS guys, one Navy SEAL serving with them as an exchange officer, an intelligence officer, and two other Green Berets to go up against about six hundred Taliban and Al-Qaeda fighters.

The odds are overwhelming. I take solace in knowing that like me, these SF and SBS soldiers have gone through a rigorous and extensive selection, assessment, and training regimen. If I have any hope of surviving, it will be because of their exceptional training and professionalism.

We're able to gain a foothold in the main gate area. It has sufficient shelter from direct fire and gives us an opportunity to try to communicate with our Afghan counterparts about the situation on the ground.

We can't stay in this spot. Anytime we try to peek around a corner, AK rounds whistle by us. Fortunately, having spent a week living here, I know the lay of the land.

I leave the SBS guys on the rooftop. They directly engage the bad guys below with machine guns as the rest of us exit the fortress and make our way to the north side, to another parapet—the last known location, I'm told, of Dave Tyson.

The steep walls and the moat around the fortress's exterior are designed to repel people from trying to enter. Thankfully the moat is dry due to drought, but the challenge now is to get up the walls and then over the top of the parapet. It's very steep. Luckily, a few of the Afghans inside the parapet take off their scarves, tie them together, and then throw it over to use as a rope.

I manage to reach the parapet. Tyson is no longer here.

There is one bit of good news. From this position, I can see both the southern and northern parts of the courtyard. The vantage point will allow me to call in and direct air strikes into the southern courtyard, where most of the resistance is taking place.

I hear screaming, see Taliban fighters charging into the northern section. Other heavily armed fighters are trying to escape the fortress, to head out into the nearby villages. The SBS soldiers use their heavy machine guns on them and break up the attack.

Now I'm faced with a "Sophie's Choice" moment.

Calling in an air strike will potentially kill Americans who are alive and sheltering in place somewhere inside these buildings. If I don't call in the air strike, I'll risk being overrun by superior numbers and allow the enemy to achieve its objective of seizing the fort. In addition to the lives lost, including possibly my own, such a sudden reversal of fortune will likely convince some of the Taliban's erstwhile partners here to take up arms again, setting back our efforts by months—or worse.

I decide to call in the air strike. There's no other choice. If I don't, we're all going to die.

* * *

F/A 18 Super Hornets equipped with two-thousand-pound bombs called Joint Direct Attack Munition, or JDAM, take off from an aircraft carrier stationed in the Arabian Sea. Using the grid coordinates derived from a map, a GPS system will guide each bomb to its intended target.

The problem is that each JDAM will strike within a "danger close" range. For a two-thousand-pound JDAM, there's more than a 50 percent probability that when the bomb detonates, it will kill anyone within five hundred meters. All of us are within that blast radius.

The first JDAM is launched—and hits short of its target, destroying about a dozen cars and trucks near the center of the compound. Seven more follow over the course of the afternoon and well into the evening, and all but one—which malfunctioned and landed behind us!—strike the desired target. We put down a lot of bombs in the southern courtyard, but the situation is far from being under control. The battle is still unresolved. And we still haven't found either Dave Tyson or Mike Spann.

Now the sun is going down, and we don't have our NODs, or night observation devices. This morning, in broad daylight and in our haste to get to the battle, nobody thought to grab our night vision devices.

We need to withdraw and pull back to our headquarters at the Turkish School. I don't want it to get overrun by anyone who escaped, or by someone who has decided to take up arms again. But first, I want to make one last search for Tyson, who I've learned is possibly inside the northern courtyard's main building.

The CIA officer with us, Glenn, accompanies me. When we reach the building, we find a gaggle of Afghan civilians and a few Northern Alliance fighters who have not gone to Kunduz, but there's no sign of Tyson. The Afghans tell us that Tyson, whom they call Baba-Daoud, had "gone over the parapet," but we're not exactly sure what that means. At any rate, the winter sunlight is fading fast, along with my opportunity to get everyone else out safely.

When we regroup back at the schoolhouse, I learn that CIA operative Dave Tyson is alive.

It turns out that Dave had, in fact, "gone over the parapet." He tumbled down the side of the fortress and then made it to a main road, where he actually got into an Afghan taxi. Because he's fluent in Uzbek and Dari, he was able to get the driver to take him back to the Turkish School, our main headquarters in Mazar-e Sharif.

I also get word that a platoon from 10th Mountain will be flying (by helicopter) into the nearby airfield. I send a contingent of vehicles to pick them up. Word of the uprising has spread rapidly, leaving all of the Northern Alliance forces left in and around Mazar-e Sharif on a razor's edge. In the darkness, they shoot at anything that moves. Somehow, my contingent manages to get in and out of the airport and back to the Turkish School without getting hit by any direct fire.

It's a chaotic night.

All of us get together and devise a plan. I grab about an hour of sleep, and at first light, we move out and reenter the fortress through a different pathway. A squad of 10th Mountain acts as our QRF—quick reaction force.

Accompanied by our Special Forces–qualified communications officer, Anthony Jarret, I make my way to the parapet on the wall dividing the center of the fortress, now even closer to the enemy.

My SBS team, an Air Force combat controller, and a couple of others are now in the opposite corner. I'm communicating with them, and they're communicating with the aircraft overhead to call in the first JDAM. Our target is the Pink Schoolhouse, which is now being used by the insurgents as a makeshift bunker/bomb shelter.

Yesterday, danger close was less than two hundred meters from the point of impact. Today, it's vastly different. Potentially deadlier for all involved. Danger close is going to be between fifty and seventy-five meters.

I hear "thirty-second countdown" over the radio and hunker down, fingers in ears and mouth open (to equalize pressure, therefore reducing internal damage), wishing we were wearing helmets. Suddenly, there is a massive blast, and the fortress is engulfed in a cloud of brown dust. I can tell the bomb didn't hit the intended target but, due to the impenetrable dust, I can't tell exactly where it did hit.

I try to radio my second-in-command, Captain Paul Syverson. No answer. I can't get in contact with him or anyone else. I get this sick feeling they were caught in the air strike.

Maybe they can't hear me because the strike was so close that their eardrums burst — or maybe they got their bells rung from the blast. I can't see them through the dust cloud and have no idea where they are.

Sure enough, the JDAM missed the intended target. Instead, it hit one of the T-55 tanks and blew it into the air, where it flipped before crashing back down, with the chassis landing on top of the separated turret. The bomb also blew a hundred-foot-wide hole through the side of the fortress.

I call in my Quick Reaction Force (QRF). We start to find the survivors. They're coated head-to-toe in brown dust. The only things we can see are their blood-red eyes and the blood dropping from their noses and ears. All of them are disoriented and in varying levels of shock.

Miraculously, every single one of them has survived. Unbelievably, there are no secondary explosions from the rockets and artillery stored just below the location where the bomb hit.

Still, I've got nine seriously injured soldiers. I call for a medevac and order a withdrawal to the Turkish School. Despite the chaos, confusion, and dangers, we manage to get everyone back to the school. It's going to take at least two and a half hours for the helicopters to arrive from Karshi-Khanabad.

It's hard to explain what being caught inside the proximity of a blast from a JDAM does to the human body. Even though the survivors haven't suffered any penetrating wounds or amputations, the shock waves are traumatic and can cause substantial internal damage and stress on organs. While waiting for medical evacuation, one of our Special Forces captains expires three times; each time he's successfully revived by our battalion surgeon, who starts his heart again and again and again.

This morning, I had fourteen people enter the fortress. Nine of them are now wounded and incapacitated, and the Taliban are still in control of half of the fortress. The mission remains unchanged: we still have to take back the fortress from the remaining fighters and find Mike Spann.

I receive some good news. US Air Force AC-130s—heavily armed gunships equipped with a fearsome variety of precision ground-attack weapons—have arrived in Karshi-Khanabad and are going to be pressed into service. Their arrival is not a guarantee of victory, but it is a welcome addition to our arsenal.

After seeing our wounded off, the remaining members of my team and my QRF head out under cover of darkness. With two CIA officers and another SF soldier, we carefully infiltrate the fortress. We begin to direct the AC-130s.

Because the gunships were pressed into service so quickly, they haven't had time to "tweak" their guns—basically, to zero their weapons. Their strikes on the southern courtyard are still effective but not nearly as accurate as they might have been.

The AC-130 strikes alert the enemy to our presence. Even though the enemy has not figured out *exactly* where we're hiding, they've got a good general sense of where we are, and we start taking mortar fire on the ground.

Unlike day one, I have remembered to bring my NOD. I can hear the mortar exiting its tube, and I can see the mortar's trail on its upward trajectory, but I can't pinpoint the enemy's locations—and neither can the AC-130s flying overhead. The first gunship expends all its rounds. A second gunship arrives on station.

Mortar rounds land all around us for hours.

The optical sights on the gunship are the equivalent of looking through a soda straw. It is a narrow field of view, but the second gunship happens to be in the right place at the right time and sees what's happening on the ground. The enemy is prepping the mortar rounds inside a building. They run outside to fire, then they bolt back inside the building to prep more rounds.

"Put some one-oh-five rounds on that building," I tell the AC-130s.

The AC-130s main gun is a 105-millimeter cannon, but the rounds they fire don't have delayed fuses, only point detonating fuses, meaning they explode on contact. They were ineffective against the steel-reinforced concrete of the Pink Schoolhouse but this building is different: the roof is made of wood.

The gunship adjusts fire and hits the building—and within about thirty seconds we start getting secondary explosions inside the fortress. The building, as it turns out, is stacked with arms and ammo, now cooking off due to the intense heat generated by the direct hit.

The interior of the fortress has now become a danger zone—especially for the Taliban and Al-Qaeda hunkering down in the southern compound. We withdraw back to the safety of the schoolhouse and begin preparing for day three of the battle.

There's more good news. Some of the Northern Alliance forces sent to Kunduz have begun to trickle back to us. Kunduz didn't turn into a gigantic battle because most of the Taliban ended up surrendering.

Now more numbers of our Afghan counterparts are back at Qala-i-Jangi and taking a more active role in trying to root out the remaining fighters in the southern compound. As they try to clear rooms, they encounter suicide bombers—guys equipped with grenades. They blow themselves up, taking some of the Northern Alliance members with them.

The fighting rages on until we end up gaining control of the southern compound. The remaining insurgents flee to the basement of Pink Schoolhouse. Unbeknown to us, there's also a tunnel that connects to one of the mortar positions.

Hundreds of bodies belonging to the Taliban and the Northern Alliance are scattered all over the fortress. Mike Spann's remains are located, right where he fell on the first day, where all of it started. We evacuate his remains back to the schoolhouse for repatriation to the US.

The International Red Cross shows up to start collecting remains, but a Red Cross volunteer entering the tunnel is quickly shot and killed by one of the remaining insurgents. The enemy and their wounded comrades are holed up in the basement of the Pink Schoolhouse. Their numbers and capabilities are unknown, but they refuse to leave—and they're still willing to fight. Clearing the dark labyrinth is a very tall order, but our options are limited.

Then one of the Afghans has an epiphany: fill the basement with water and freeze them out.

It's an excellent idea—genius, to be honest. The daily temperatures have been somewhere in the thirties. At night, it gets very cold. The remaining survivors won't be able to withstand freezing cold water for long.

Flooding the basement proves to be the decisive factor. After several hours of standing in chest-deep, freezing cold water, the remaining prisoners surrender.

One is a Taliban fighter who speaks English. The others call him the "Irishman." The man in question turns out not to be from Afghanistan but from California.

His name is John Walker Lindh, a twenty-year-old who converted to Islam when he was a teenager. He went to school in Yemen then traveled to Afghanistan so he could help the Taliban fight the Northern Alliance.

To this day, I remain convinced that the six hundred–plus Taliban and Al-Qaeda fighters who initially surrendered did so as part of a plan—a Trojan horse to get them inside Qala-i-Jangi. We didn't have time to teach our Afghan counterparts how to properly handle prisoners of war—and the enemy knew this. The prisoners had been poorly searched, which allowed them to smuggle arms and ammunition and stage an uprising to take control of the fortress and its impressive cache of armaments.

But the enemy hadn't anticipated the devastating effects of US air strikes or the ability of our Special Forces to fight in a dense and somewhat urban environment. Out of the six hundred fighters who arrived at the fortress, only eighty-five survived. The numbers were insignificant to the Taliban, but the lost opportunity to recapture Mazar-e Sharif was a devastating blow.

There were multiple opportunities for things to go horribly wrong for us, but we managed to recover and persevere, get through the battle with only suffering a single casualty, CIA operative Johnny "Mike" Spann.

Years later, when I'm working as a Director for Counterterrorism on the National Security Council in the Obama White House, I'm involved with a controversial matter involving Private First Class Bowe Bergdahl, a twenty-three-year-old Army soldier who deserted his military outpost in southeastern Afghanistan and, within hours, was captured by the Taliban. He is released by the Taliban in exchange for five Taliban senior leaders being paroled to Qatar from Guantánamo Bay. The exchange is controversial — to say the least!

Bergdahl is a knucklehead and did a lot of things wrong. Plus, there was lots of Taliban propaganda to the effect that Bowe had voluntarily switched sides. Even though the tales were false, I'm not entirely surprised when people around me say, "Screw him. We should just leave him behind, let the enemy have him."

My response is always the same. "I would never want to be part of an organization that makes a post-facto decision to leave someone behind — to decide that some soldiers are not worthy of being recovered. No matter what happens, no matter the circumstances, we don't do that."

The commitment that soldiers make to the men and women on our right and left is, "Hey, I'm not going to let you down. I won't leave you behind." We're always fighting for our own lives, but we're also fighting for the lives of our fellow soldiers and for those we've lost on the battlefield—Americans like Mike Spann, the first casualty of the war in Afghanistan.

When I think back to the battles at Qala-i-Jangi, I'm always reminded of one of the best-known lines from Thomas Babington Macaulay's epic poem "Horatius at the Bridge":

Then out spake brave Horatius,
The Captain of the gate:
'To every man upon this earth
Death cometh soon or late.
And how can man die better
Than facing fearful odds,
For the ashes of his fathers,
And the temples of his Gods,

We don't fight for idols or temples but for our families and the ideals enshrined in our Constitution and the Declaration of Independence: that all men are created equal, and governments derive their just powers from the consent of the people. I think about the men who were at my side during those days, how I took a great deal of comfort being surrounded by them. How, during those battles, we all felt like Horatius guarding the bridge against overwhelming odds—and emerged victorious. Together.

Brian M. Kitching

Captain, US Army
Conflict/Era: War on Terrorism (Afghanistan)
Action Date: October 4, 2012
Silver Star

It's the middle of the night. My brother Julian and I are sitting in my beat-up car, talking about our futures. I'm grappling with how to find more purpose through leadership and service.

"If that's what you're really interested in," my brother tells me, "you might want to consider joining the Army."

It's what he really wants to do, but he's working through some health challenges. (Julian will go on to have a distinguished career as a Green Beret.)

I've recently started college in Huntsville, Alabama. That night, I decide to drop out and enlist.

"You're throwing your life away," friends and family say. "You're going down the wrong path."

I do it anyway.

AMERICAN HEROES

My first duty station is Fort Campbell, located on the Kentucky-Tennessee border and home to the Army's storied 101st Airborne Division. My report date is September 11, 2000.

I don't know anything about how the Army works. I'm the first in my family to serve. I didn't have anyone in my life who could coach me on so many of the dynamics unique to the Army and military writ large.

But I instantly connect with Army life, and I'm fascinated by the stories of people from all over the US and world who have volunteered to serve our country. I'm also fortunate to have mentors who encourage and challenge me to always care for people and strive for excellence.

It's not long before we all face a challenge that changes everything.

One morning, I'm putting my camouflage uniform on as I'm walking down the hallway in the barracks. I look over at a TV and see two big office towers on fire.

"Is this a movie?" I ask the other soldiers in the room. It's September 11, 2001.

Leaders begin yelling, *"Go to the arms room and draw your M4!"* Everyone is grabbing their rucksacks and weapons and scrambling to secure different areas of Fort Campbell. In the following days and weeks, security at Fort Campbell tightens dramatically, and our training schedule intensifies.

Over the course of the next several months, we receive information slowly—and train relentlessly. Our infantry platoon sergeant has us stand in formation while he reads aloud accounts from combat in Afghanistan. For someone who has just joined the Army, I struggle to imagine what combat looks like halfway around the world.

In March 2002, I deploy to Afghanistan as our brigade finishes wrapping up Operation Anaconda. I'm a corporal, a forward observer for 3rd Battalion, 187th Infantry Regiment (Rakkasans). Essentially, my job is to conduct intelligence activities and call for indirect fire from mortars, close combat aviation, or in some cases, fighter jets. It's unclear how long we'll be in-country.

Almost no one in our formation has any substantive combat experience. Many of us assume that we'll be facing direct combat the moment we hit the ground.

That's not what happens. When we land at Kandahar Airfield, the forward operating base (FOB) is still in the process of being built out. There are minimal combat patrols on that first deployment.

Our days are filled with various work details, rehearsing for potential missions, and pulling guard duty for detainees. We train constantly, and sit around our tents talking about home, or whether we'll see action.

One day, General John "Jack" Keane addresses our formation. "Make no mistake, we're going to be over here for a long time."

"No way, we're going to be out of here in no time," mumbles one of the young officers standing near me.

A few months into the deployment, I'm notified that I'd been accepted into the Army's Green to Gold program. It selects enlisted members to complete college and become commissioned officers.

Over the next decade, I finish college as an Infantry officer, and after completing Ranger School, I deploy to Afghanistan three more times as a platoon leader—once with the 82nd Airborne Division, and twice with the 75th Ranger Regiment, the Army's premier raid force.

I take command of a mechanized infantry company (essentially, an infantry company that delivers soldiers via the Bradley fighting vehicle) in November 2011. Only a few weeks later, I'm notified that we'll be deploying to Afghanistan. We'll be relying principally on our feet to do the job.

We deploy in March 2012. This will be the first Afghanistan deployment for most of the soldiers and leaders in the company. Most of these soldiers are brand-new to the Army, kids who have only qualified on their weapons in basic training. Many of the leaders have several combat experiences from Iraq, but this combat will be different.

Most of them have relied heavily on vehicles to conduct operations—and there isn't a strong culture of dismounted operations. Up to this point, I've always prioritized combat-focused training in the organizations I led, but when I learn of the specific area where we will be in Afghanistan, it gives me pause. I know I'll have to intensify our efforts.

The district of Panjwai in Kandahar Province is notorious. It's one of the birthplaces of the Taliban and known for fierce, well-equipped fighters and elaborate improvised explosive device (IED) tactics. It's also one of the most violent areas in the entire country.

Generally, I think it's more effective to spend time cultivating a greater sense of buy-in within a team, but we don't have the luxury. In the ninety days leading up to the deployment, we push our soldiers hard, emphasizing extended foot movements, functional fitness, nutrition, and patrolling tactics. We study our enemy's tactics and regularly review real-time reporting from the unit we'll be replacing.

"This is the way we must train in order to meet the demands of combat," I tell them.

Our company operates out of Combat Outpost (COP) Sperwan Ghar, a towering, man-made piece of terrain in Panjwai District, Kandahar Province, west of Kandahar City. I'm one of three people in the 135-man unit with previous combat experience in Afghanistan. A company of Afghan National Army (ANA) soldiers is colocated with us.

Our mission is to prevent the Taliban's ability to destabilize Kandahar City with large-scale attacks. That means finding and destroying weapons, explosives, and fighters attempting to organize operations against our partnered Afghan forces.

We quickly begin conducting operations in our area. Within weeks, we destabilize the Taliban's operations, knocking out key IED and drug facilities. We patrol day and night. We're disciplined with after-action reviews, and we're continuously refining our tactics based on the operational environment. The noncommissioned officers and young soldiers of the company perform with astonishing bravery every day.

But our operational successes come at significant cost. By September, we've lost five men, with more than a dozen wounded.

Our company is scheduled to redeploy in December, and the fighting hasn't let up.

In late September, I'm given the objective to conduct a company helicopter assault in a village named Nejat, to clear and destroy the village's IED facilities. Nejat is by far the most dangerous village in our area of operations — notorious for running drugs and weapons. Its tight clusters of mud structures and mazelike paths make it a nightmare for IED attacks. Of the two hundred–plus battles in my area of operations over the course of the nine-month deployment, almost forty have occurred in Nejat alone.

We infil the company on CH-47 helicopters on October 3, 2012, and I spend the evening with 3rd Platoon, clearing the east part of the village. Second platoon is running logistics for us and resupplying as we clear the area on foot. By the time night falls, we've engaged in several firefights. I grab a couple hours' sleep on the dirt next to my radio telephone operator (RTO).

The next morning, on October 4, we link up with 1st Platoon to finish clearing the last group of buildings in the east while 3rd Platoon moves to clear the southern portion of the village.

Because of the IEDs in the area, we go to extreme lengths to limit the enemy's ability to predict our approach. I insist we vary the design of our movement formations and routes to get into the villages. We travel in single rows (files), often dispersed, sometimes climbing mud walls, or using axes and picks to tear them down. But it's incredibly hard, slow work. Brutal. Covering eight hundred or so meters can take hours.

On October 4, we move in a file, even though there're thirty of us, plus some ANA soldiers. What's also taking time is that at the front of the formation, I have someone searching for IEDs with a dual-sensor handheld detector. It uses ground-penetrating radar and metal detection technology that can locate IEDs even if they're buried a bit deeper into the ground. The clearing device isn't great, but most of the time, it's better than nothing. It's the only one 1st Platoon has, which creates an additional risk for us.

The day is heating up when I start receiving reports through my RTO about intercepted radio traffic. The Taliban is watching us and is planning an attack.

Many times, it's a bluff meant to scare us off. But in this area, anytime I receive those reports, there's always been a fight.

We move into an area of the village with a maze of sharp turns and alleyways, and disperse to clear a series of reported IED facilities. It's the afternoon, and I can tell the men are exhausted, hot, and nearly out of water.

Nejat is relatively quiet. When we get to an intersection, my RTO reports that the Taliban is close enough to hear us.

"Freeze," I tell the fire team. "Do *not* move."

I have thirty people with me; one wrong move and we could potentially trigger an explosion.

All our "elements"—individual squads, platoons, and other units—are spread out. They won't be able to support, and the high chance of friendly fire in this area would be too risky. Not to mention all the IEDs.

I pull a set of graphics from my pocket so I can orient myself.

As I look to the north, an AK fires at us from about fifty to seventy meters away.

We return fire and look for cover.

When we break back to the rest of the formation, I direct one of the team leaders to use an M320 Grenade Launcher Module (GLM) to fire multiple grenades onto the enemy's position.

Reports say there are five to seven fighters. After ten minutes of fighting, the enemy breaks contact.

Staff Sergeant D, a phenomenal NCO and one of the fittest guys in the company, is operating the mine-clearing device. The day is incredibly hot. When we approach a gate leading to the final sets of compounds, a lot of people are low on water.

The staff sergeant stops by the gate and says, "I'm getting a high reading."

"Let's use one of the line charges," I say. We worked with a nearby Special Forces detachment and made line charges from detonation cord, or "det cord"—thin and flexible plastic tubes filled with explosives—attached to C4. It's much more efficient than carrying around the heavier device that launches line charges. We throw a line charge onto the path and, sure enough, two IEDs detonate.

We move around the gate, into a huge marijuana field that's around two hundred by two hundred meters. The plants are massive, probably twelve feet tall, and provide us some concealment, but no cover. And it's hard to stay quiet when you have thirty guys moving single file, trying to avoid IEDs.

We're probably midway through the field when we start

getting hammered by heavy machine gun fire from a PKM, a Soviet-manufactured machine gun with a muzzle velocity of seven hundred meters per second.

Everyone drops to the ground as rounds cut down the plants and leaves around us. I'm in the middle of the formation, screaming at everyone to return fire. No one does. They're physically and emotionally exhausted and possibly suffering heat exhaustion, disoriented from dehydration. Most of them are kids. While they're trying—wanting—to do the right thing, their capacity is somewhat limited.

Two people have been wounded—one of my squad leaders and an ANA soldier. At the moment, there's very little I can do to affect the tactical situation while stuck in the middle of a marijuana field. I need to get to the front.

I emerge from the field and see a few soldiers being treated for injuries. My RTO and I find cover behind a small dirt wall and receive reports on the wounded. The rest of the platoon begins returning fire.

I direct my RTO to initiate the call for the medevac aircraft. We start to return fire. We're pinned down in an open area, and we need help. I call for attack helicopter support.

A pair of Apaches arrive and identify ten to twelve fighters moving amid a maze of paths and firing at our position with PKMs, AK-47s, and RPGs. Our helicopters engage enemy positions while we wait for the medevac to arrive.

The helicopters begin firing and launching rockets into the village. I mark the landing site for the medevac with smoke, but the helicopter flies directly over the landing zone and lands in an adjacent field—with an eight-foot wall between us.

Why didn't they land in the LZ? Did they see something that scared them? Made them nervous?

There's no way we can move our wounded over that wall in the middle of a firefight. "Stay here," I tell my RTO, and break into a run.

It's so hot. I'm dripping sweat.

I run about fifty yards and find a break in the wall large enough to go through. I sprint to the helicopter and direct the pilot to go back to the area we marked. It's not far, but the helicopter's got to go over the wall.

As the pilot moves over the wall, the helicopter starts taking gunfire from various positions. As we return fire, I'm hoping and praying the enemy doesn't launch an RPG at the helicopter.

We're only able to get one of the casualties on the medevac helicopter before it's forced to take off. The other wounded will have to continue the mission. There's no way that medevac helicopter is coming back.

The enemy gunfire was coming from a cluster of compounds about one hundred meters away. I see huge flames leaping into the air from a large pile of trash and straw that got hit by a rocket from one of our attack helicopters.

I do a water check. These kids barely have any. I can tell they're all smoked.

We need to move forward and continue our mission of

clearing the rest of the village. We still have adequate ammunition and two Apaches in the area.

We move up to a wall. This one is ten feet high. I climb it and drop into the first compound. After I establish security, I tell Staff Sergeant D to follow.

"I want you to establish security east of our position," I say, taking the mine-clearing device from him. "I'm going to clear the courtyard of this structure so we can get the rest of the platoon safely out of this open area." I'm thinking, *I don't really know how to use this thing, but I've got to demonstrate some courage and confidence for these soldiers if we're going to get this done and get out safely.*

When I'm relatively certain the area is safe, I give the go-ahead and the rest of the platoon starts flowing into the courtyard. We take some pop shots from a fighter with an AK-47, but we're somewhat protected by the surrounding walls.

Staff Sergeant D, now back to working the mine-clearing device, says he's getting readings for potential IEDs.

My RTO receives a report. "The Taliban wants to bring in 'more friends and the big gun.'" He says this in front of most of the soldiers. I wish he hadn't.

I'm pretty sure it's just a tactic, but the fighters in Nejat are well equipped. These kids are terrified. I can see it in their eyes.

I've spent more time in Nejat than anyone else on the ground. I've moved multiple platoons through this village and know the area inside and out.

There's an open area that leads to a wall a hundred yards

away, directly to our front. Behind that wall is the final grouping of structures we need to clear.

It's no one's job but yours to get your men to safety, I tell myself. *No one else will be able to do this.*

"I'm going to make that run," I say to one of my teams. "Cover me, and I'll secure the other side."

It's the longest run of my life.

I make it to the other side, start pulling security, and we're able to bring the rest of the platoon up.

"We're getting too many readings," the platoon leader says, meaning he assesses there are a lot more IEDs somewhere around us. "I don't know what we should do."

My soldiers have been fighting most of the day. Some are close to becoming heat casualties. No one wants to move. Out of the hundreds of patrols I've been on, I've never seen this level of fatigue—but we can't stay put.

I clear the last few structures on my own, attempting to complete our mission. The Afghan Army soldiers refuse to assist. I report to my battalion commander that we're complete with the objective. I tell my soldiers we're going to move to the exfil point outside the village.

Given the IED threat, this will be an extremely risky movement.

The only way we're going to make it out of here is if I lead us out. These men have fought so hard and sacrificed so much this deployment. I go to the front of the formation and tell my men that I'm going to lead the way. Staff Sergeant D will follow from a safe distance. We have only the one mine-clearing device, and my soldiers will need it if I get killed leading them to the exfil point—and there's a strong chance I will.

I take a moment to mentally say goodbye to my wife and son. I love them so much.

I know this area of the village well. I walk north as the point man, choosing my path carefully where the dirt is packed almost like concrete, and keeping my weapon at the ready should enemy fighters emerge from one of the alleyways. My heart pounds with every step.

As I skirt around a crumbling wall, multiple IEDs detonate, one directly in front of me and another one directly behind me.

I'm blown into a wall. When I regain awareness, all I can see is dirt and dust swirling, pebbles falling around me. Then I hear people yelling. I get on the radio, and as I start asking for reports, I look up and see the mine-clearing device, broken in half, suspended in a tree.

I don't see Staff Sergeant D anywhere.

We find him lying face down in a nearby creek. One leg is gone, the other mangled. Blood is running downstream. My RTO and a couple of other guys jump in the creek to retrieve the wounded staff sergeant.

I send the casualty report over the radio, and the platoon medic, in the rear of the formation, moves up to respond.

As he does, he triggers another IED. It costs him his leg.

Another sergeant is wounded in the face from shrapnel. He's covered in so much blood it's indescribable.

It's a nightmare. The most violent deployment I've ever experienced.

Before long, the medevac helicopter is inbound. The pilot radios to ask if the field has been cleared.

"There's no way we can clear it," I say. "This casualty needs to be evacuated. He will die if you don't come and get him."

It's an extremely tight space, but they land. I carry the staff sergeant to the helicopter with a few other soldiers. He's barely conscious as we place him on the litter, but he grabs my arm so fiercely out of pain I can feel his fingernails piercing my skin.

In that terrible moment, I pause to look at him. I touch his head softly, hoping with everything in me that he'll be okay. His remaining leg is just sort of dangling there. There's no question he's going to lose it.

The medevac aircraft takes off. I consolidate our remaining forces and begin moving. As we reach the exfil point, I'm devastated — for our wounded, and for the soldiers that remain.

But I keep reminding myself that everyone is still alive, and that we did the best we could with what we knew and the resources at our disposal. Combat feels natural to me. And while I don't enjoy some of what I've experienced in battle, there's nothing like fighting side by side with my soldiers. The most unbelievably magnetic experience is witnessing what soldiers will do to save the lives of their friends.

That's what motivated me that day in Nejat.

Conrad Begaye

Staff Sergeant, US Army
Conflict/Era: War on Terrorism (Afghanistan)
Date of Action: November 9, 2007
Silver Star

We've just finished our meeting with local leaders in Afghanistan's Nuristan Province and the fastest way home is to take the trail the 10th Mountain Division calls "Ambush Alley."

Home is Forward Operating Base Bella, in northern Afghanistan. The Hindu Kush Mountains here look like a *National Geographic* photo spread. Just beautiful. This place could be a tourist attraction if it weren't for the people trying to kill us.

I'm Navajo, born and raised on a reservation in Shiprock, New Mexico. My parents moved us to Tucson, Arizona, where my uncle lives. He and my other uncle served in Vietnam. One was a tanker, the other in Special Forces. I joined the Army because of them and got an Airborne Ranger contract.

I'm up front, leading the patrol across a little mountain trail full of trees. The right side is straight-up mountain. The left is nothing but pure cliff. I hear a gunshot, and it immediately brings me back to a moment four years ago, in 2003. I had just arrived in Iraq when I heard these whizzing sounds. *What in the world is that noise?* That sound, I quickly learned, was made by bullets whizzing past me.

I turn around and see soldiers taking on fire, including the Marine who's joined our patrol. The platoon leader is fatally shot in the head and drops to the ground.

Someone pushes me. When I turn around, I don't see anyone. I feel that push again and again but each time I turn there's no one there. That's when I realize I'm taking on rounds that are being absorbed by my body armor.

We're taking fire from all directions, and we're shooting in all directions. An RPG slams into the mountain and rocks the ground beneath my feet. More RPGs start coming in, and the enemy gunfire isn't letting up. It's insane, a full-bore onslaught, one that's been well planned.

Justin, one of my gunners, runs up to me and yells, "They're all around us."

"Where they at?"

"I don't know."

I'm afraid, but I've trained for this. I've been in several firefights, one of which lasted for three days. But this one is different. Here, we're pinned down. Here, we have mountains behind us and steep cliffs in front of us. There's nowhere to run or hide —

Something pinches my shooting arm. I keep firing.

"Dude," Justin says. "You're shot."

I look at my arm, see that it's bleeding. I quickly wrap it up and get back into the fight. My gun jams. There's no way to clear it, so I toss it aside and grab one of the many rifles scattered across the ground.

The amount of fire we're taking on is overwhelming. I need to get these boys out of here, but how?

An idea comes to mind.

No, I can't, I tell myself, even though I know it's the only way out. We don't have any other options.

"Everyone," I call out. "*Follow me.*"

I jump off the cliff, hugging the side of the trail.

I tumble against the rocky slope, taking on fire, and come to an abrupt stop when I get caught in a tree.

Other soldiers follow suit and jump. Corporal Sean Langevin, my SAW gunner, gets caught in the tree, too. He's so close I hear him say, "I'm shot."

"Brother, we're all shot."

As we're untangling ourselves from the tree, Sean is shot again. He falls down the side of the mountain, and all I can think about is how Sean, one of our heroes, is scheduled to leave Afghanistan soon to go home for the birth of his baby girl.

When I finally get to stable ground, I discover that Justin has also been shot. I take him to this little cave I've spotted and lie on top of him so no one can shoot him—but plenty of rounds are landing inside the cave, hitting my body armor and plates.

One paratrooper has been shot in both legs. The bad guys are still firing at him.

"Play dead," I call out to him.

The enemy will direct their fire elsewhere if they think he's dead. It works.

I stay with Justin until nightfall.

"Hey," I tell him. "Let's put our NODs on and see what the boys are doing."

The squad has fifteen guys. When Justin and I rendezvous with the others who've taken cover behind rocks farther down the mountain, some of the squad is still MIA. I check my radio, amazed to find that it's still working.

I call for medevac. I'm told it will be at least an hour because the helicopters are in Baghdad. Someone gets on the horn to the base in Kandahar. Fighter planes are sent to assist us.

"I don't need fast fliers," I say. "I need helicopters. I need Apaches." They're the best option because we're in the mountains.

A Spectre gunship shows up. The woman on the other end of the line tells me that the remaining members of my squad are roughly thirty feet from my position.

There's no way to climb back up there. It's too steep.

Another one of our guys—our medic—is MIA. He's recently engaged and expecting a baby. Because he's MIA, the gunship won't shoot on our position, but it's tracking all the warm bodies moving on the ground—including the enemy fighters. The woman I speak with tells me where they're located. I use the information to call in mortar fire.

"The Taliban," she says, "they're all bugging out."

She means they're leaving. *Oh, thank God.*

We retrieve Sean's body and carry it to the medevac point. Our medic who was initially MIA, I discover, has been killed. Five soldiers and one Marine are dead, and the flight medic is stacking bodies on top of each other.

"Hey," I say. "Show some respect."

"They're dead."

"I don't care. Don't stack them on top of each other like that."

My armor plates have turned to mush, but other than the gunshot wound and some bruises, I feel okay. I'm considered walking wounded because of the round I took in my shooting arm, so I get patched up quickly and am flown east to the airfield in Jalalabad.

While I'm lying in my hospital bed, recuperating from surgery, I'm told I suffered a traumatic brain injury. A general comes in and pins me with a Purple Heart.

I'm sent for a month of treatment with the Wounded Warrior Unit in Ramstein, Germany, before moving on to the Army station in Vicenza, Italy. Every three months, I travel back to Ramstein to get CT scans.

One day I'm walking down to the chow hall when some guy approaches me.

"Hey, where you going?"

"To get some chow," I say.

"Your ceremony is happening."

"What are you talking about?"

"They're giving you the Silver Star."

I had no idea.

I don't tell people I'm a Silver Star or twice a Purple Heart recipient. I don't really talk about what happened that day. It's a bad memory for a lot of people. I keep it to myself. Put it away in a bottle.

McKenna "Frank" Miller

Sergeant, US Army
Conflict/Era: War on Terrorism (Afghanistan)
Date of Action: December 17, 2010
Silver Star

I fly in as one of the only passengers on a C-130 packed with cargo, and after landing at the sprawling Bagram Air Base, I'm picked up by our supply sergeant and taken via 1980s Land Cruiser to our unit logistical cell. The *entire* twelve-man Special Forces team is consolidated, either sitting in or on our gun trucks or pacing around smoking cigarettes (which we never did outside combat). These are familiar faces, but their expressions worry me. Most of the guys look disheveled, almost resembling soldiers in an *Apocalypse Now* scene. Distraught.

Forty-eight hours ago, I was on R&R in the Canary Islands and stranded due to a volcano eruption in Iceland that grounded most flights throughout Europe. I was given a short break before my deployment — my reward after having graduated from a three-month stint at one of Special Forces' more demanding intelligence courses—but a rather direct message from the commander makes it clear that I need to meet my team in Afghanistan immediately.

Our commander is a very animated and energetic major and has called a meeting in one of the standard, plywood B-huts. The moment I step inside, he starts asking questions about the Tagab Valley. The lush valley an hour's drive from Kabul is known for its bountiful pomegranate harvests—and its constant fighting.

Later I learn why my ODA—Operation Delta Alpha—is so disgruntled. We've lost good people from our unit in this valley and in the weeks they have been on the ground, there have been limited opportunities to seek out and send the enemy back into their mountain hiding areas.

I'm answering the commander's questions about how to stabilize the region when he says, "All right, that's what I thought. You're the team sergeant for the Tagab team."

"What?" My sense is he wants a more aggressive approach to breaking the Taliban stronghold, but I wasn't expecting this assignment.

One of my only mentors throughout my military career was a team sergeant in the Tagab Valley. On November 10, 2007, he was gravely wounded with one of his men, SFC Pat Kutschbach killed, during one firefight. Prior to the deployment, I was unhappy when I was told that I would be going to another region that is considered less contested. Now I realize the calculus has changed in terms of operations but not in how I do my job.

The team, is now going to be divided in two, or "split team." Half of us will be in Tagab. The other half will go to Surobi, another district of Kabul Province. That valley isn't friendly, either. French soldiers recently suffered a massive attack that resulted in a bloodbath.

I'm assigned to Tagab. My team and I will be partnering with the legendary French Foreign Legion mountain troops.

Rumors have swirled around the French Army's Foreign Legion since its founding in 1831. People think its ranks of 8,000 are swelling with convicted criminals or people looking to escape their pasts. While there is some truth to that, the Foreign Legion is a highly professional organization, and the mountain troops are some of the Legion's best.

Initially, the relationship between my ODA and the French leadership was not as productive as I would have liked. Their officers are highly educated and seem to apply campaign planning to whatever it is they do and can be fairly rigid on deviating from plans that their higher command have approved. The new team leader, Captain "Dave" Fox, and I meet with the French general to discuss how we can work together to establish a local Afghan defense capability. The NATO coalition needs to develop and support these local guys, and the way we're going to do that is to build formidable checkpoints throughout the valley manned by Afghans from those villages. This local security concept has been used since the early days of the war, but the Afghan government, understandably, wanted to evolve into professional security forces.

Captain "Dave" Fox and I both knew that history has proven

local "militia"-type security forces are the only effective means of local security in tribal cultures. If the locals can resist the urge of abusing power and becoming an enemy themselves, the villages can defend themselves. My point to the French general was that the Afghans must accept this as their fight. We're just here to help or assist, we will eventually leave, and I am only going to put my ODA at risk for operationally relevant threats. We use the local forces as the first line of defense but use the hammer of the ODA if we must. That's the mission of Special Forces.

I believe our ODA and the French developed a decent relationship because our nations understand irregular warfare and began to appreciate the methodology we were implementing. When one of our vehicles gets shot up or disabled, they send out roughly a Coy (or Company) of infantry, armor, and heavy recovery vehicles so we can continue fighting while the French work on recovering our vehicles. We help them out on their operations by coordinating close air support, moving in small elements to close with the enemy, and amass massive amounts of firepower on threats. They seem to trust us, and we trust them.

The French build incredibly efficient and structurally sound bunkers like no other. They'll build towers out of wood to create fighting positions. The goal of the checkpoints was to provide a defendable fighting position should they become overwhelmed, and my ODA and the French provide indirect fire, like mortars or artillery while the Afghan security elements hunker down while we work on eliminating the threat, or enemy.

We needed to establish one last checkpoint to complete the

defensive ring we have worked to develop over the past six months. This last checkpoint will be the most difficult to construct due to the terrain and the requirement of a fighting position at the crest of a mountain, and due to the small roads, heavy equipment will be limited. This does not take into account that this particular area is quite volatile and enemy forces can amass quickly by coming down from the upper valley. Because of these considerations, we decide that a thorough site assessment will need to be conducted to ensure we have the appropriate designs and request the appropriate materials for the actual construction. It takes weeks if not months to receive some construction supplies, so I need to make sure we got the construction right on the first try and as quickly as possible.

For this assessment mission, I'll be taking my ODA, Romanian special forces, and Afghan fighters who are loyal to the government. I ask the French if they could assist us with this last checkpoint, and they surprisingly agree to support. For the assessment, they would send us three senior engineers to make sure this checkpoint doesn't end up falling off the side of the mountain.

"We'll take some measurements and come right back," I tell them. "If something goes down, just find me or another American. Whatever happens, we'll deal with it. Just stick with us and everything will be okay."

Our main checkpoint, we've decided, will be in the small village of Jalokhel, which is about two and a half miles as the crow flies from our camp, but the commute takes around 30 to 45 minutes without fighting. Even with the expected fight we would have during the mission, this would be a fairly easy task. If, we did not have to climb the hill and do measurements there. The height of the fighting position would rest around forty-six hundred feet in elevation, which means we would have to hump roughly three hundred vertical feet with over slope angles of 27 to 35 degrees. Not terrible for a leisurely hike, but adding seventy-five-plus pounds of kit and getting shot at makes this patrol more of an event than a hike. We need to control the high ground in order to protect the checkpoint, which will be on the valley floor.

As I finish our patrol brief, I go back into our tactical operations center (TOC) to make sure our command in Kabul or any of our intel services have any updates. Our 18F (Special Forces Intelligence Sergeant) states that FBI local informants are reporting a "big weapon" in the vicinity of Jalokhel Village. We always received warnings about enemy activity, but this was the first time informants had said something like "big weapon," which isn't very helpful.

I am thinking the enemy had managed to get a DshK heavy machine gun or recoilless rifle into the area, which is a decent threat, but I have state-of-the-art optics, snipers, close air support (CAS), and an observation capability that can cover the entire valley. Dave and I make the decision to execute.

In the Tagab, we have limited use for our monstrous Mine Resistant Ambush Protected Vehicles (MRAP). I believe you lose most of your situational awareness and maneuverability using these wheeled fortresses and it is a process to just get in and out of these things as opposed to our completely open gun trucks.

I had to negotiate with the Special Operations Command Europe (SOCEUR) Commanding General (LTG Mike Repass) when he recently came to visit our camp. He lost his mind when he found out I was still using our gun trucks, which he outlawed due to vulnerabilities. I argued that by using appropriate tactics and elementary maneuver doctrine, my ODA is actually *safer* using lighter armored gun trucks.

On board our gun trucks, we typically have a Mini-Gun, which the Taliban call "the Breath of Allah," three M240s, a 60 mm mortar, three 40 mm grenade launchers, an M249 "Saw" machine gun, half a dozen shoulder-fired rockets, and around ten thousand rounds of various types of ammunition.

While the MRAPs can accurately identify and engage targets up to two miles away, the vehicles are so heavy they crush the irrigation systems and damage walls during tight turns in villages, which pisses off the locals. So, I position the MRAPs where they can overwatch my gun trucks and foot movements and use the MRAPS to relay radio transmissions and act as our Quick Response Force (QRF). The CG (commanding general) reluctantly agreed to allow my team to use the gun trucks for the remainder of the deployment, but we will cease to use them anywhere else.

As we turn off the main road and into the outer portion of town, radio chatter on our scanners picks up. There is really only one route to where we need to go and the enemy knows that. We can hear the bad guys saying they're getting into position. About five hundred meters out, we start taking fire.

These conditions are typical, so I establish a defensive perimeter and then get the support guys and engineers to dismount and send the Romanian snipers to their preselected positions. "Go up to the top of the hill and do what you need to do," I tell them. "We'll provide covering fire. Don't dick around and take all day. Get back and we'll return to base."

Team leader Dave Fox takes off hiking up the mountain with the three French engineers, two of our local Afghan security guys, and the Romanian ODA team leader.

Several of my security positions report about ten military aged males (MAM) doing the typical "spray and pray" from concealed locations. We've positioned our vehicles at angles to absorb some fire but far enough away where gunfire is less effective and enemy rockets that actually hit my trucks may skip off or detonate into the ground.

My vehicle crew are some of the finest men I have ever served with and we are truly a family. I feel as if we are the definition of brothers from another mother. My gunner, a six four, 230-pound chiseled 18 Bravo (SF Weapons Sergeant) who is the spitting image of Michael Jordan and nicknamed Baby Jesus because he was born on Christmas, is letting the Mini-Gun rip with its characteristic *wwwwwwamp*. My equally monstrous 18D (SF Medical Sergeant) is on the strong side of our truck and dumping rounds from his M240 as my 18E (SF Communications Sergeant) is on the weak side but coordinating indirect mortar fire with the French back at Kutschbach.

The enemy is concentrating their fire on us as opposed to the element climbing the hill—which is exactly what I want because the diversion has allowed our Romanian brothers to infiltrate the mountainside with snipers. Their hide (site) is situated on the NE side of Jalokhel mountain roughly three hundred meters to my left, neutralizing any threats they identify. We also have air cover to keep the enemy in check.

An hour or so into the fight, Baby Jesus gets shot in the turret. The wound isn't bad. In fact, he didn't even know he had been hit until he noticed blood soaking his uniform — he'll survive — but I'm more concerned about the battle. Things are starting to unwind a little bit. I get on the radio and call Fox.

"Dave, we need to wrap this up — the sooner, the better."

The enemy starts shooting RPGs, recoilless rockets, and mortars.

Everywhere I hear explosions — including one from the top of the mountain. Being a demolition guy, I know that explosion doesn't sound right — a deep thud, like it came from somewhere underground — and it's accompanied by a lot of dust and dirt. I get on the radio and call Dave again.

He doesn't respond.

I try again and again and get nothing. I call the Romanian sniper element.

"What's going on? Was that an RPG explosion on the top of the mountain?"

"We're not sure," they reply.

I need to get up there and figure out —

"Urgent...surgical."

Dave's voice.

"Urgent...surgical," Dave says again. I turn to our medic, thinking how I should have sent him up with Dave and the others but did not want to take him off the gun. My initial thought was we bring any injured down and he treats them in a less exposed area. In the meantime, we need as much firepower toward the enemy as possible to facilitate extraction from the ridgeline.

I call out to Matt, my other 18B, who is bouncing around from position to position, checking weapon status and coordinating with other security positions when I yell at him to get up the mountain and find the recce team. Matt is a fitness God and is notorious for wearing the bare minimum in kit and usually sporting some type of modern hiking or running shoes. While on patrols, I tell him, he is not allowed to wear the "five-fingered" footwear popular with cross-fit types. Matt can hump the mountain no problem and do it fast and could carry a truck down if he needed to.

"Matt, you and I need to get up there and see what the hell's going on."

I move the trucks a bit because they're taking on a lot of fire. Rich, our support radio guy on loan from Kabul HQ, starts physically dragging Afghans into position to provide cover fire while Matt and I start humping up the mountain.

It felt like a near-vertical ascent and we're taking on an ass-ton of enemy fire. I recall having to move farther to the NW, or the backside of the mountain, to limit round impacts near me.

As I approach what I think was the summit, I see two of the French engineers dragging their captain down the mountain. As they notice me approaching them, their faces — faces I will see for the rest of my life — tell me everything I need to know.

Oh, shit, he's fucked. Then I realize that the captain's taken a non-survivable impact to the head.

Above me and to the west, Matt is carrying our legendary local security commander, a guy with a commanding presence despite weighing one hundred pounds soaking wet who has an arm and a leg just barely attached by some muscle and ligaments.

The entire party, I find out later, was standing on top of a buried IED when it detonated. The explosion launched the entire recce team airborne. They land piled together, the dead and the injured.

Medevac is incoming. Matt carries our security commander down a more westerly route as our Air Force JTAC is trying to bring a medevac into the riverbed below the mountain and obviously less exposure to accurate enemy fire. My team leader Dave is half crawling and walking across down the mountain toward me and the French element. The French engineers... they're pale and clearly shell-shocked. They will not leave their dead leader—one of them carrying the captain's exploded weapon—and not making much progress and stumbling due to the massive overpressure absorbed by their bodies. Their world just got rocked. They're non-mission capable and out of the fight.

"Leave the weapon," I tell them as I cut the kit off the captain and attempt to check for any signs of life still left in him. "Leave all this other shit and take his plates (body armor) and helmet—*and get down.*"

I throw the two-hundred-pound or so French captain over my shoulder. Dave is about 90 percent non-mission capable because he just got blown up. I give him my rifle and say, "I can't shoot, so provide cover fire."

Dave is totally out of it. "Damn, your weapon is so light."

"Thanks, man. Let's go. Just cover me."

The vertical climb up the mountain was difficult. Going back down while carrying someone over my shoulder, under enemy fire, with zero cover—it's going to be treacherous. What I don't know is that the enemy force has grown. Instead of dealing with ten bad guys, I'm now dealing with roughly seventy according to intelligence reports after the fight.

Dave returns fire as we make our way down the uneven terrain. I keep falling and getting back up, as enemy rounds impact around us. As we approach the more NE portion of the mountain, enemy fire became more accurate as we are more exposed. It feels as if I am being pushed forward, and after some confusion, I realize spalling from rocks and rounds are ricocheting with some impacting the first man I have lost during a fight. Later, I realize the heavy emotional thought that the captain may have saved my life despite his life leaking from his injuries all over me.

Once I reach lower elevation, I begin to feel the physical impact on my body. During a previous fight, I had been sprinting with ammunition from one truck to another and strained, possibly tore, a couple of muscles in my right hamstring complex. Now, as I try to stand up after yet another fall on the rocky ground, I feel a fresh tear.

The pain takes my breath away and I fall again onto the rocks.

I fight my way down, dodging enemy fire. When I reach the bottom of the mountain, I am pushed down again and as I try to stand again, tear the other hamstring and collapse.

There's a riverbed, or wadi, nearby. It's December, not quite rainy season, so the water level is low. Still, the channel offers cover and concealment other than being below the angle of direct enemy fire. Once I reach it, I'll then have to navigate approximately two hundred meters across the wadi to where the trucks are providing cover fire.

Dave and I make it another fifteen or twenty feet. Another nearby explosion knocks me down but I try to pick everything back up and keep going.

The Afghans providing cover fire don't come out to help. Dave tries to change that. He runs ahead of me. I can't remember how many times I fell and got back up. I make it to the edge of the riverbed but there is a ledge from the washout I need to climb up and over. The next thing I know, Rich, my radio guy, and the Afghans are pulling myself, Dave, and the French captain out.

There's a tree nearby. They lean me against it, and I feel my body shutting down. I start tuning everything out. Someone gives me water.

Holy shit, I made it.

Then I start hearing the gunfire and explosions and my mindset changes.

You're a leader. You've got to turn it back on, man.

I'm in pain and realize my body is exhausted, but my body and mind go into another physiological mode. As long as I can still think and make coherent decisions, I'm going to fight and lead. The minute I can't, I need to be man enough to take myself out because I'll get someone hurt or killed.

I can still fight. I can still think, so I can still lead. My guys, these people — the French and the Afghans — they're all relying on sound leadership.

I get on the radio, trying to find out where everyone is. I'm thinking particularly about Matt, whom I have not seen since catching a glimpse of him at the crest of the mountain. Matt was carrying a man down — I remember that. I remember the medevac trying to fly in multiple times, and I remember seeing sparks from enemy rounds coming off the helicopter. I remember that brave crew possibly making more than one attempt to land under fire. I believe they would have tried to land if we would have asked them to, but our Air Force Special Operations controller and my 18E, who is also a qualified JTAC, skillfully called an audible and moved the HLZ, or helicopter landing zone. Matt and Dave had performed adequate lifesaving procedures on our Afghan brother and the condition of the French captain was not going to worsen.

Matt checks in — he's alive and coordinating loading injured onto a Light Medium Tactical Vehicle (LMTV) to move the casualties south to the main road where the medevac has a better chance of not being shot out of the sky. Once the injured are loaded onto the LMTV, it moves west for the bumpy and tight route to the HLZ. Matt provides security during the transit.

Once I get accountability of all our forces, I collapse all of our security and defensive positions. I am certain the enemy have made it within our north and south flanks of the route to the helicopter landing zone (HLZ) on the main road. I decide to remain dismounted to engage any threats. Perhaps I am so full of adrenaline I do not want to remove myself from the reality of battle you sense when on foot.

Baby Jesus is in the turret of my truck following behind me and asks me if I am hit.

I reply, "No man, I'm good."

"You're covered in blood."

My 18D, who is in the rear of my truck, comes on the radio and asks in his normal angry tone, "Frank, are you hit?"

"No, it must be from the casualty." I then realize that I have failed my men in giving them updates during the evacuation of the casualties. Some of them had no idea to the extent of the casualties, much less that we have a KIA. I can feel my medic's emotion over the radio. I have made a mistake. I have one of the best medics on all of the teams, but because I chose to take myself out of a leadership role and assist with casualties, I could have cost someone their precious life by not pulling my medic from his gun and letting him save lives.

To get to the LZ, we're going to have to drive roughly four hundred meters through the town. We're pretty cool with the inhabitants, but you never know.

The enemy has indeed moved on one of our flanks. They are now firing sporadically from the left or south side of the road. The enemy parallel us as we start a convoy, a few of us running in front of the trucks so we can provide cover fire or, even better, take down the enemy.

We reach the HLZ. My 18F and others are already helping load the injured onto the helicopter and it departs moments after the main element arrives. Dave decides to stay in the fight and refuses to be medevaced and mounts up back into his gun truck.

We notify the French we are returning to base. I do not recall a word being spoken during the short drive back to the camp. We pull through the gates, all the French are staring at us as they have been able to witness the entire battle from standing on the walls of the camp.

Despite the fight, I do not deviate from our nonnegotiable standard operating procedure (SOP). I help with refueling the trucks, then direct everyone to shovel 3-6' empty brass covering the truck floors, empty rocket tubes, ammo cans, expended 40 mm grenade shells, look for any bullet holes in the fuel/brake lines or engine, and clean all heavy weapons before chow.

I find Dave and we make the silent walk from our vehicles to the French Tactical Operations Center (TOC). I don't realize I'm still in my blood-soaked kit. This is what they'll see when we tell them they have lost an officer and debrief on the operation. French soldiers continue to stare at us. I feel they are staring in anger, holding me directly responsible for the loss of one of their respected leaders, despite my telling him, hours ago, everything would be fine.

Good people just don't quit. They stay in the fight and lead. They run through a situation knowing there's a fifty-fifty chance of taking one in the head because they want to do their job but also protect their friends and be a teammate. Good people always go above and beyond. Good people never leave anyone behind. Leaders never quit.

Neil Prakash

First Lieutenant (Armor), US Army
Conflict/Era: War on Terrorism (Iraq)
Date of Action: June 24, 2004
Silver Star

Night missions are the worst. Nothing ever happens during the 3:00 to 6:00 a.m. shift. Or hardly ever.

The four of us crew members inside our M1 Abrams tank are listening to music, courtesy of the MP3 player we've wired into the comm system. Tank commander (that's me); Sergeant Joshua Pritsolas, my gunner; John Wayne Langford, my loader; and Johnathan Mewborn, my driver, patrol the area where we're setting up observation posts. Baqubah is an Iraqi city northeast of Baghdad that's known for its date and fruit orchards.

Our tanks always did patrol missions in pairs. After we secure a route, we park back-to-back, about five hundred meters apart, on opposite sides of the road. My gunner is scanning for hotspots, and I'm looking through my binos and chewing Red

Man when an IED explodes between the two tanks. The 155 mm round, I'll find out later, was hooked up to an egg timer.

So much for my logic about the route being secured. Fortunately, no one is hurt.

At 6:00 a.m., we head back to Forward Operating Base Scunion. As I'm refueling, I hear a little bit of machine gun fire in the distance. I call it in, and then head back out for my seven to ten o'clock shift.

The gunfire increases. It's not sporadic—it's a hellacious amount. To date, I've never heard that much gunfire. High-intensity conflict is definitely going on, maybe one to five kilometers from us. We can't see it, but it sounds like it's coming from the city.

I call it in. I'm told 1-6 FA (Field Artillery) Battalion is in contact.

For the next three hours, I receive reports about groups of five to twenty-five men armed with RPGs and AK-47s moving to my location. I'm not worried. I've been hearing those same reports for the five months I've been in Iraq, and not a single one ever came to fruition.

Still, I relay the reports to my wingman, Staff Sergeant Kenneth Terry. A group of five to twenty-five dudes is going to be hard for the lookout to miss. Plus, we're surrounded by farmland, so there's no way these guys are going to sneak up on us.

At 10:00 a.m., I'm told: return to the forward operating base (FOB) and report to your command post. When I arrive, I see the XO, two other platoon leaders, and their platoon sergeants gathered up. I'm wondering what's going on when my company commander comes in with a map.

"Here's the deal," he tells the group. "Alpha Company is going to exit the FOB. We're going into Baqubah. Enemy fighters are already in there. The mortar platoon has set up a blocking position at the traffic circle. We're going to pass through them and then enter the city and engage the enemy. First Platoon will be in the lead. I'll be behind them."

Everyone mounts up. We're at REDCON-1. Full alert, ready to move and fight.

We head out with four tanks, a Humvee, the M113 APCs (armored personnel carriers), and another four Abrams tanks in the rear. Our command sergeant major stands at the gate to see us off.

I'm in the lead tank. We move together in a column formation. The company commander calls with a SITREP—a report on the current military situation in the area.

"We are looking at roughly 250 fighters in the city. They are arranged in groups of four. They are going to initiate contact with IEDs. This will be followed by RPGs and heavy machine guns and AK-47s. They will also use hand grenades. The attack will come from the left and then it will come from both sides. They are in the streets and on the rooftops and in the alleys, and they are moving through the alleys with Bongo trucks."

I have never received detailed enemy SITREP like this before.

Oh, shit, we're heading straight into a fight.

The city's buildings are about two to three stories high. At 11 o'clock, I see an old factory-looking building about one thousand meters away. An RPG flies from somewhere inside, heading straight at us.

"Scan left," I tell Sergeant Pritsolas. "RPG."

He sees the smoke trail as the RPG flies straight past us. I fire the main gun and hit the building. The round leaves a hole maybe ten feet wide and keeps going through several walls.

Langford takes a few seconds to load another main gun round.

"*UP!*" he shouts, then makes himself as thin and flat as possible against the left side of the tank. When the cannon fires, the breach goes back, and it will break his legs if he's in the way.

The median, loaded with twenty-three IEDs, explodes in a daisy chain. RPGs are coming at us from every direction — from the rooftops and alleys and second-story balconies. They're firing two types: RPG-7 high-explosive (HE) rounds and RPG-7 armor-piercing (AP) rounds.

When HE rounds hit the tank, the tank doesn't move. The explosives just spatter and leave a mark like a black soot snowball. AP rounds — the penetrators — are a different story. When one hits, the tank shudders.

The HE rounds make a lot of smoke when they explode. I can taste the smoke because my hatch isn't fully closed. We never operate fully closed. With a fully closed hatch, I won't have situational awareness. We're operating open hatch, my head's out, and I'm eating all that smoke and inhaling the smell of cordite.

I see machine guns. Star-shaped muzzle flashes are going off everywhere. Even thirty to fifty meters away, I'm thinking, *God, I hope I don't get hit in the face* while taking aim and engaging them with my M2 Browning .50-cal heavy machine gun, better known as the "Ma Deuce."

Then I see guys with grenades trying to get close to us. They're going to try to lob grenades into the hatches.

"We need to go to open protected mode," I announce over the net.

Open protected mode is where the hatch is popped up just enough for me to look out. A fighter could easily shove his muzzle through this small opening, but first he must get on my tank. I give the open protected mode command not only to my platoon but also to my commander.

We're moving real slow now because we're trying to engage the fighters—and they're everywhere: a hundred guys in front of us and on our sides; on the roofs; the alleys; and they're all shooting at us. Taking out our tank would be a huge strategic win for them.

"Stop," I tell Mewborn. "I just saw guys down an alley."

The gunner has controls that allow him to slew the tank left and right. He can also fire a laser that will come back and tell the tank the range. Then the tank's ballistic computer calculates elevation.

"Where are these guys?" Sergeant Pritsolas asks. "I don't see them."

"I got it," I reply.

I have a control interface similar to the gunner's, the command turret override; it allows me to snap to targets since the gunner has a narrow field of vision and the tank commander usually has a 360-degree field of view. I take control of what looks like a giant aviation joystick, squeeze the Cadillac gauges, and swing the turret—with the 7.62-mm coaxial machine gun armed—as fast as the turn will go, which is very fast.

I see ten, maybe fifteen guys dressed in white robes, their heads covered in red and green scarves—and spray them down. I grab the hand mic that's set on the company net and report, *"They're everywhere. They're on the rooftops. They're in the second stories, the streets and alleys, everywhere."*

I tell Mewborn we need to back up. I peek behind me to see what the gap is between me and Staff Sergeant Terry.

That's when I realize we're all alone. The second tank that was accompanying us is gone.

"Sir," Mewborn says, "just took a direct hit with an RPG to my hatch."

"You alright?"

"Yeah. My periscopes are shattered, I can't see. I don't know how much more she can take."

I'm amused. Mewborn is my tank driver and he used to be a cop. But I'm certain none of us are material science engineers, and I'm laughing inside as he's calculating the number of RPGs his driver's hatch may or may not be able to withstand.

"Roger, Mewborn. Move out."

It's controlled chaos inside the tank. The Nuclear Biological Chemical (NBC) Warfare alarm is screaming in my ears—in everyone's ears—through our Combat Vehicle Crewman (CVC) headsets.

The alarm is a new one for me. I don't even know something can be done about it.

"*Sir,*" my gunner yells, "*hit the NBC alarm.*"

I look on the control panel to my right and see the NBC

alarm is flashing a red light. I turn it off. The alarm stops, then goes off again.

I can't get it to turn off. Later, we'll learn the alarm was triggered because the NBC filter system mounted on the side of the tank took a direct hit by an RPG.

As we continue to drive, I'm hoping one of these RPGs doesn't hit the rear sprocket. That's the drive sprocket; it's the only thing that moves the tank track and makes the thing go. If we lose it, we'll be what's called a "mobility kills," which means the tank goes from being an asset to a liability.

We'll be sitting ducks.

The tank, thankfully, is continuing to drive, thanks to the brilliant men and women who designed it at the US Army Tank Automotive Research, Development and Engineering Center (TARDEC). We need to stay mobile.

I get on the 50-cal gun. It's manual and mounted on what's called a cupula. While it can rotate 360 degrees, it can't stay in a fixed position because my gunner is in the turret, traversing left and right, hitting targets, scanning for targets, hitting targets. If I turn west and my gunner turns east, my .50-cal keeps wanting to go with him.

This isn't marksmanship. I'm just trying to hit dudes that are close enough that I'm pretty sure I'll get them at the rate of fire. It's a shit show.

A guy appears about three hundred meters in front of our tank. He grabs an RPG, and as he slings it over his shoulder, I can see his eyes and the green scarf wrapped around his head.

He's standing in front of a dumpster and I'm thinking, *The balls on this guy, squaring off with an Abrams tank.*

Our main gun fires. The flash obscures everything in front of me. When the smoke clears, the dumpster looks like a blooming onion and the bad guy is no longer there. But more fighters are coming.

"Sir," my gunner says. "We got a turret malfunction."

I grab the override.

My gunner says, "I can't turn the turret."

"Let me try."

The turret is frozen. While it can't move, the gun tube is still able to elevate up and down, allowing us to still find targets.

We keep engaging the enemy. The temperature inside the tank is 136 degrees. I know because the ammo temperature gauge says so and Sergeant Pritsolas needed to input that data into the ballistic computer before we got into this fight. The alarm is still blaring, I'm exhausted and thirsty, and I'm reminded of a speech the sergeant major gave us at Ranger School during the summer of 2003.

"You are in possession of dangerous knowledge, which is you know better than anyone else just how far you can push yourself. Other men may want to break along the way, but you won't let them because you carry knowledge that no one can take away from you."

I can get through this. I've been in tougher spots. I can get through this. We're going to be okay. We'll get through this.

I contact the commander and recommend that White Six, the platoon leader for 2nd Platoon, come up.

"He's immobile," the commander replies.

White Six, it turns out, broke from the column formation, in an effort to flank the enemy. He ended up sliding his tank down a ditch and sheared the sprocket. Now he's stuck.

"You need to get White Six and recover him back to the FOB."

My new task is to find White Six's tank, hook it up to mine, pull him out, and get both of our tanks to FOB Scunion for recovery.

My wingman comes up behind us. "Red Six, this is Red Eight. Your tank is on fire."

"Negative, Red Eight."

As we're moving, I get another message. "Red Six, this is Red Nine, your tank is on fire."

I investigate and find the problem.

The RPGs that hit us didn't do any damage to the tank's armor, but they *did* blow up my sponson boxes. Made of thin steel, these storage boxes, located on the left and ride side of my tank, hold rifles plus first aid kits, all the technical manuals, Gatorade, MREs, and assault packs—*that* is what's on fire.

I arrive at the parking lot of a different building and find White Six. The tank has slipped a bit down the ditch, and because it's at an angle I can see that the rear sprocket is sheared off.

I drive up to him and grab our tow bar. As I drag his tank back to the FOB, all I care about is getting my tank functional again, rearming, and getting back into the fight.

* * *

My gunner and loader are all rearming while a bunch of mechanics look over our tank.

The tank has what's called a "bitch plate." It's made of metal, weighs about one hundred pounds, and covers the engine pack. One of the RPGs hit it, and a piece of steel is wedged underneath the turret.

Mechanics use massive, seven-foot-long pry bars to loosen the steel. When they finally free it up, the turret is operational.

I contact my commander. "We're coming back out."

"Roger," he replies, and gives me the company's position.

When I arrive, two rows of two tanks are parked side by side. Our orders now are to get to the Twin Bridge from the west. A mechanized infantry unit will approach from the east and secure the far side of the bridge while we secure it from the east—and secure the city.

Along the way, the brigade commander tells us insurgents have taken the governor's family hostage at the civic center, a five-story building.

"You are main gun free to engage any hostiles and fire from the civic center."

We don't take on any more fire. The other weird thing is that we see no dead bodies. The fighters have recovered all the corpses.

By the time we get to the bridge, my guys are complaining about how thirsty they are. I decide to get some IVs going and, remembering my Ranger School training, tell them, "You guys are fine. We're going to be okay."

We don't take on any more contact. As day turns to night, I'm expecting another attack, but it never comes. When the morning of June 25 dawns, we return to the FOB, high on enthusiasm and victory.

The next day, as the mechanics examine my tank, I find out that it took the brunt of all the strikes. One RPG blew off a headlight and left a large hole in the front slope. An armor-piercing round hit the front slope of the turret.

My gunner's face was behind that armor.

One of the mechanics is curious at how deep it is and sticks a long piece of soldering rod inside the hole. "The AP round," he says, "was close to penetrating the inside of your turret."

If it had penetrated, it would have fucked up my crew.

There's another hole in the side of the depleted uranium armored skirt — and another hole on the opposite side. We do a little investigating and discover the round lodged in the track section. If it had gone a few inches lower, it would have missed the track and kept going into the driver's compartment and killed my driver.

We got lucky.

My battalion commander puts me in for a Silver Star, one of the military's highest honors. I tell him I'm going to put all my guys who were in the tank with me up for the award. Every single one of them deserves it.

"Absolutely not," he says. "You may pick two guys to put in for ARCOM V's."

An Army Commendation Medal is a mid-level award given by local commanders to soldiers who exhibited consistent acts of heroism and meritorious service. The "V" device, as it's called, denotes acts of heroism involving conflict with an armed enemy.

"That's not right," I say. "They all—"

"You can pick two. I'm not doing blanket awards."

I've always respected the New England Patriots' simple motto in their locker room—"Do your job." My guys did their jobs perfectly. Flawlessly. I didn't do anything more than them. It was just moving, shooting, and communicating.

This is bullshit.

I try to convince my battalion commander that my platoon deserves awards with a higher precedence than just two ARCOMs with V's. The battalion commander isn't swayed. He's made his decision—and it breaks my heart.

Tim Sheehy

Lieutenant, US Navy
Conflict/Era: War on Terrorism (Afghanistan)
Date of Action: April 9, 2012
Bronze Star Medal with Valor device/Purple Heart Medal

I've decided to resign from the Naval Academy. I'm eighteen years old and halfway through my plebe year.

Coming to the Naval Academy has been my plan for a decade. I was the biggest aviation dork. My neighbor was a Korean War Navy pilot. I started flying planes with him when I was eight. I even went to space camp, and 9/11 only furthered my desire to serve.

I hand in my resignation letter to my company commander. It's December 2004. The country has been in Iraq for a year and a half. In the last couple of weeks, we've had our first Naval Academy folks killed in combat since Vietnam.

The company commander stares at me like I've lost my mind. "Really?" he asks. "Are you serious?"

AMERICAN HEROES

"I want to quit so I can enlist and fight."

The truth is, I don't fit in here. Like every kid, I saw *Top Gun* and wanted to be Maverick, but now I don't identify with the brash Navy pilots and their frat-boy style. The people I like—the people I see myself emulating and working with in the future—are the solemn tribe of war fighters who are bearing the brunt of what would end up being America's longest war. More specifically, the most highly trained and battle-tested warriors in our nation's legions—Special Operations, Navy SEALs, Army Rangers, Marine Recon, Green Berets. That's my DNA. I wanted to be where I could be most impactful for our country, and the door kickers were it. All of the interactions I had had at the academy to that point only affirmed this.

"You're the top guy in the class," he says. "Military, physical, academic—you're crushing it across the board, and you're telling me you want to quit so you can—what?—become a private in the Army and just walk the streets of Baghdad?"

"The country is running two ground wars. Me sitting here going to school and learning how to be a jet fighter pilot or a submarine driver—the country doesn't need that. Our country needs guys on the ground, so I'm going to go where I'm needed most."

"It's your decision. I'll get to work on this. In the meantime, I'd like you to meet someone."

The next day, I go to Dahlgren Hall to meet with a guy named McDonnell. He's wearing an Army uniform, what's called the "bus driver outfit"—green pants, a green shirt, and green coat.

Holy shit. This guy's a Green Beret and a full-bird Special Forces colonel.

"I heard through the grapevine that you're thinking of quitting the Naval Academy and joining the Army or Marines."

"Yes, sir."

"I respect that. Take a seat."

He tells me he's going to the National Defense University in DC, after having returned from his most recent Iraq deployment. He did a double-pump deployment in Afghanistan and also served in the first Gulf War and in Panama and nearly everywhere else.

"I can tell you right now that I don't need any more privates kicking doors," he says. "What I need—what the country needs—is young leaders. Strategic decisions are happening at the lieutenant/captain level. That's where this war will be won and lost.

"The Navy SEALS are, in my opinion, one of the best if not the best special operations organization. They're tough and incredibly tactically capable. They have capabilities no other group has, but they have an organizational disadvantage because they speak a different language, plan a different way, and don't share a common professional development track with their special-ops peers."

In the first few years of the war, there had been a lot of intertribal challenges between the SEALs and their Army counterparts Rangers, Green Berets, Delta, and others. "The other units mesh better together because they plan the same way. They have the same rank structure and speak the same language and they all derive from the same professional development pipeline."

I don't see the Army adapting to the SEALs, so the SEALs better figure out how to adapt to the Army.

"We're exploring this new experimental program," he says. "Sometimes we'd send SEAL officers to Ranger School so they could get to know the Army. The problem now is with op-tempo. No one has the time, people get injured—not a viable path. What we've decided is that the best place to achieve this common tactical language proficiency is Ranger School."

Is he offering me some sort of job? If so, what's the catch?

"We've already talked to the superintendent," McDonnell says. "He's open to letting us grab a handful of midshipmen and have them do this as a trial course. From what I hear, you're a top performer headed to BUD/S, you want to do it?"

"Fuck yeah, are you kidding me?" (At least that's what I said in my head... What I actually uttered was probably more like "ummmm... Yes, sir, Colonel, sir....")

"Great. The deal is, when you finish this program, you need to graduate and become a SEAL officer."

I became the first midshipman ever to go to Ranger School. I learn how to lead from infantry officers, Ranger officers, and SF officers. I attend multiple other schools and get a chance to deploy.

Four days after I graduate from the Naval Academy, I report to Naval Amphibious base Coronado, the Naval Special Warfare Center for Basic Underwater Demolition/Seal (BUD/S) training. I make it through all three phases and graduate in 2009, at the height of both wars, and get my Trident. I'm twenty-three.

Five days later, I'm in Iraq.

* * *

I have some decent Arabic under my belt, so my basic function is as an interim translator and liaison between Army units and the SEAL team stationed in Baghdad conducting hostage rescue and high-value target elimination.

I'm hot with excitement to get downrange and finally do what I was trained to do. That's what's going through my head and the heads of every single guy here who's around my age. We've spent months — or, in my case, years — getting ready for this opportunity. I'm motivated, full of piss and vinegar to finally get to do the job that I've wanted since I was five.

There's a dividing line in the military between guys who have been in-theater and guys who haven't. Even if you're the smartest, strongest, fastest, coolest guy in the world, if you haven't been in-theater, no one gives a shit about what you have to say. Shut up and get in the corner.

Nobody wants to be that guy, especially someone like me, who at some point in the near future is going to be taking over a unit. I need the experience and legitimacy of "having been there."

I do five months in Iraq. I return home in April of 2010, and forty-five days later, I leave for Afghanistan, where I learn a ton and participate in several challenging operational scenarios, including a hostage rescue mission involving a British aid worker who is rumored to also be a British intelligence agent. I'm tasked with helping find her, and I'm given access to a bunch of experimental aerial assets.

We find her. Unfortunately, she's killed during the rescue mission.

AMERICAN HEROES

The next time I deploy to Afghanistan, I'm in command of a SEAL platoon. We arrive in January of 2012. This deployment, our job is very different than during my previous deployments. Instead of high-value targets (HVTs) and hostage missions, our job is to live in remote Afghan villages and teach them to protect themselves. In typical military euphemisms... it's called Village Stability Operations. We live in mud huts alongside the people to help make them self-sustainable, from a security aspect, so they won't get rolled over every time the Taliban comes to town.

The Taliban are bad guys, but they're more of a passive mafia element that's enabling and making money off groups like the Haqqani Network and Al-Qaeda (which eventually become parts of ISIS). These groups smuggle weapons and drugs into the villages, terrorize the people, steal food, and take boy children from their families to be soldiers for their militias.

We experience combat nearly every day, sometimes heavy, sometimes light—but the enemy is omnipresent, and since we don't have a FOB, we are always "outside the wire." A lot of my men—and myself—get wounded, and while it's a challenging year, we make some fantastic gains. We teach them to stand up to the local Afghan police and how to defend the village and not be subject to Taliban Haqqani and Al-Qaeda pressure.

It's a great mission. I love it far more than I ever imagined I would. And while I love the impact we're having in the area, the gains we have made have been felt by the enemy leadership in the region. Resistance is getting stiffer and the networks we are fighting are getting more aggressive.

On one of our patrols to a neighboring village to set up a

checkpoint there, we start hearing enemy radio chatter. In code, but an indication that something is coming. This is nothing new, we hear this all the time. But what's different this time is it seems they have a spotter within our village. They know too much about our route and our team.

One of my guys listening to the ICOM channels tells me someone is giving the bad guys our exact location. "He's buried IEDs. They're gonna hit us with rockets from the ridge, push us into the IED field to the south."

The spotter network is always our biggest enemy. Spotters hide in the general population. Trying to distill who is good and who is bad is the worst part; but finding and catching them is the best part because it makes a significant impact.

"Damn, get the DF equipment up and let's figure out where he is — I want him —"

Shooooops! Bam!

We start taking fire from the ridge.

I've been fighting nearly every day for months and months and my brain goes into combat mode. Any combat vet will tell you war brain kicks in. Sometimes it's a good thing — I can get into a calm, Zen-like state — or, in this case, it's bad because I'm beyond pissed off that someone in the village is deliberately trying to kill me and my teammates. My team is my family.

When we find the spotter, I bust into his home and, right in front of his family, grab and throw him to the floor. I press my rifle barrel against his forehead while the rest of the team starts searching the place for guns and explosives, everything. To this day, I'll never forget how the skin wrinkled on this guy's forehead, or how his children looked at me in fear and horror.

I joined to be the good guy. The hero. A force of good.

Now, to these people, I'm the bad guy. A six-foot, 230-pound bearded American Viking monster pressing a rifle barrel to their father's head. He was spotting on us, yes, but to his family he was the hero, I was the villain. What scared me most is I was ready to pull the trigger—I really was. When I realized that... I knew something was... wrong.

"There is no morality in war," the saying goes. "Morality is the privilege of those judging war from a distance." The fog of war is real. And war does change you.

I've been living here in the desert too long. I've got to remember my values.

In April of 2012, I'm serving at a remote outpost. The insurgency kind of owns this part of the countryside, but we've been capturing physical terrain, so we are expanding into a new region and establishing outposts to hold the gain.

I'm walking along a tributary, low ground between two mountains in the Arghandab River Valley with an Army contingent, SEALs, and roughly eighteen Afghanistan partners under my command when we start taking on fire from multiple locations somewhere uphill. Everyone maneuvers for cover as we figure out where the fire is coming from.

My team is nicely spread out across the soft valley soil, with a small overwatch contingent on the high ground, and they immediately start suppressive fire as we maneuver. I'm the ground force commander, and what I want now, more than anything, is to kill the ambushing element—which appears to be in some buildings along the ridgeline—and complete the mission.

There are missions where trading your life for success is important. Osama bin Laden, a hostage rescue, a nationally important target...this was not one of those. We accept death as an incidental risk to our mission, but rushing into a hail of gunfire is not it. We will break contact, flank and kill them, or maneuver to high ground and kill them with rockets. But we are not going to charge across open flat ground to attack an elevated position. I'm going to make sure my guys make it back safely.

Craig, one of the officers, radios me. "We got a wounded over here on the left flank."

Team leaders start moving their elements into a left flank maneuver. I make my way closer to Craig's position, dodging enemy fire, to assess the situation. I get within fifty meters and see he's on the radio and laying down suppressive fire with his M249 Squad Automatic Weapon (SAW). The gas-operated and air-cooled light machine gun can engage targets up to eight hundred meters away.

"Who's wounded?" I ask.

"It's me." Craig's saying it like it's no big deal even though I see he's got a tourniquet on his leg. Total stud. So proud to be with him. He's bleeding, but he's still fighting. People say millennials are weak, but I've seen Craig show amazing strength, just like he's doing now.

War brain kicks in, and this time I'm so Zen. So calm. *Craig is in a good cover spot, but it's small. If the triggerman on their PKM machine gun moves just a few feet, he can get a better angle on Craig and finish him off. If I can reach him, I can pick him up and get to that ridgeline fast enough that the enemy can't get a bead on me. If they do, the worst-case scenario is I get killed while I'm making the run.*

My biggest concern is not becoming a casualty for my guys.

I coordinate the team in effective covering fire. I come up and run through enemy fire to Craig's position.

Anytime the enemy sees a bunch of us crowding together, they suspect we're trying to help someone who's wounded. They target their fire on us while I throw Craig on my back, in a fireman's carry, and run up the rocks and through the ridgeline to get him out of the contact zone. The team lays down a heavy barrage to cover us, and although there are a few close calls as we move, we get out fast enough and within a couple minutes are behind a rock ridge next to a nice little field.

I place him against a rock wall in the field and assess Craig's condition. The bleeding is serious, 7.62 right through his thigh. I took a ricochet round during the event, but it just grazed me. Craig is rock-solid, totally with it, tourniqueting himself and offloading his gun and ammo so we can keep it in the fight. I radio Craig's location to the medic. It's tough to find us. Many people don't realize, but the low ground in Afghanistan, the "green zone" as we call it, is thick and dense. Up high on the ridges it's sandy and rocky, but as soon as you're in the loamy river soil, you're in the bush. Our medic can't find us in the thick brush. I punch out of it to go find him and bring him to

Craig. We are a small team. We need all the guns in the fight, so we didn't bring anyone off the firing line to come help out with Craig.

Craig needs the medic fast or he's going to bleed out.

I run back through the contact area. As I guide the medic to Craig's location, I call in a medevac. It will be a while. With Craig under the care of our medic, I grab his AW and ammo and head back to the team.

I drop the AW with another teammate so we have the firepower back online. The enemy is moving on us just as we are moving on them. I secure help from one of my teammates to provide covering fire, and then I push back up to high ground on the left flank. I grab the grenade launcher and four of my guys and we start gaining ground. The heavy base of fire from our team halts the enemy movement and drops a few of them. The bad guys are too high for us to successfully execute a full-flanking push, but as our AW fire pushes them back, we take out a few of them.

After we suppress contact, we start maneuvering back to the LZ. Helos are inbound. The LZ for medevac is low ground and although we killed the ambush element, they may have friends around, so my team leaders pull back a bit to the best high ground and hold there to provide covering fire for Dustoff (the call sign for the medevac birds) in case of a counterattack.

With such a small team across a wide area, we don't have a lot of people to spare. Our core elements job is clear: shoot bad guys if they try to disrupt Craig's medevac. I make my way back to Craig's position with the medic, mark the DZ with green smoke, and help carry him onto the helo. Give him a hug and a handshake as we load him on, not sure I will ever see him again. Luckily, he gets home, pushes through his physical therapy, and fights hard to get back into theater within six months. He's a great American.

People always ask what went through my head when I decided to run through gunfire. They think it's like the movies where the main character sacrifices himself by deciding to run into the firefight while this swelling, emotional music plays in the background. That never happened with me. It was pure clarity and calm decision-making.

My team is my family. Helping them — saving them — that's all that mattered.

Jay Zeamer, Jr.

Captain, US Army Air Corps
Conflict/Era: World War II
Date of Action: June 16, 1943
Medal of Honor

Interview with Barbara Zeamer, widow

I'm twenty-one years old in January of 1949 when my parents and I take a train trip from Indiana to California, where I go to school. We get as far as North Dakota when our train gets caught in a blizzard.

To pass the time, my father decides to get a card game going. "Do you play bridge?" he asks everyone in the club car. Only one person answers.

His name is Jay Zeamer and he's thirty. He and his boss are traveling to Seattle from Hartford, Connecticut. When a storm in Chicago grounded them in Minneapolis, they switched from plane to train.

Jay clearly doesn't know how to play bridge. After a couple of hands, I get the distinct feeling that Jay lied so he could meet me.

The blizzard causes a forty-eight-hour delay. We miss our connection out of Seattle and have to book a hotel. Jay's boss suggests we share a cab since, it turns out, we're all staying at the same place. That night, Jay asks me out to dinner. Just the two of us. I say yes.

Afterward, we go out dancing. The next night, we do it again. I take off Sunday morning and head to Los Angeles.

I'm back at school, at my sorority house, when my father calls from home.

"Barb, you just got a dozen roses."

"Who are they from?"

"The card says, 'To the Queen of Hearts from the Jack of Diamonds.'"

"Oh, that can only be Jay."

My father calls me again the next day. "Barb, Jay just called. He wants you to meet him in LA tomorrow night."

"I can't do that." My father knows I'm dating a fella who is in the Air National Guard and building a new home in San Bernardino. "Just forget it."

"Barb, he's already on his way."

I don't know what to do about the man I'm seeing. In the end, I decide to make up this awful story so that I can avoid him and meet Jay. He takes me to the Beverly Wilshire Hotel for cocktails. We proceed to the Coconut Grove for dinner and then we take the train to my aunt's house in Alhambra, California. When we disembark, he proposes.

I've known this man for less than a week. Including tonight, we've gone out exactly three times.

"I don't even know you," I say.

Still, there's something about him. Something that feels meant to be. The next day, I say yes.

Five months later, when we get married, we're still getting to know each other. I know Jay served in the Army during World War II, but Jay never talks about his time during the war.

I have no idea he won the Medal of Honor. I don't even know one thing about the award until one day the Army asks Jay to speak about his experiences. He goes to a theater and tells his story before a noon movie. Afterward, the Army takes us out for a meal.

The next year, we go to Washington, DC, and attend the Inaugural Ball. It's exciting. Everything in my life starts getting exciting and wonderful.

We live near Boston and have five daughters. Jackie is a junior high student when a teacher says to her, "You must be very proud of your father."

"What?" Jackie asks. She's learned a little from her friends about her father's war service, but no one has ever called him a hero.

"He's going to the White House for the Medal of Honor Society," the teacher says. "I heard it on the radio."

Jay begins sharing with our daughters stories about his student days at the Massachusetts Institute of Technology, how he fell in love with planes before enlisting in the Army and serving in World War II.

AMERICAN HEROES

In 1942, Jay was stationed in the Pacific Island state of New Guinea, where he led a team that rehabilitated an old B-17 bomber. The following year, on June 15, he volunteered to pilot the bomber on a photographic mapping mission near Buka, Solomon Islands. There, Jay spotted twenty Japanese fighters taking off. During the aerial fight, Jay was shot in the arms and legs. One leg took fire from a 20-millimeter cannon.

Slipping in and out of consciousness from blood loss, Jay continued to pilot the damaged bomber and complete the mapping run. His bombardier, Second Lieutenant Joseph Sarnoski, was fatally wounded during the attack. He shot down two planes before he collapsed and was posthumously awarded the Medal of Honor.

Jay's battle was recorded as the longest battle in military aviation history. It was also the first time that two men in the same aircraft received the Medal of Honor.

Jay's injuries landed him in the hospital for eighteen months. Still, I always had the sense that he didn't feel he deserved the award. He was so close to his bombardier, and he felt terrible about his being killed.

I'm ninety-five and don't know how much longer I'm going to be here. I have five daughters, seven grandchildren, and three great-grandchildren—including my grandson and great-grandson, both named Jay. My gift to all of them is a scrapbook on my husband's life, from when he was born until he died in 2007, at eighty-eight.

I just finished the scrapbook. This Christmas, I'm giving Jay's namesakes manila envelopes stuffed with information about him. So they'll have something to remember him by.

Earl D. Plumlee

Staff Sergeant, US Army
Conflict/Era: War on Terrorism (Afghanistan)
Date of Action: August 28, 2013
Medal of Honor

I'm a Green Beret and I feel like I'm being punished.

I've just finished serving on an Operational 1434 as the lead of the A-Team, the heart and soul of Special Forces. I trained and worked with local militias and the paramilitary Afghan National Police to eliminate the Taliban and create good governance to run the country.

Though we risked shoot-outs with the Taliban daily, it didn't take long to drive the insurgents out of town, permanently. Miri, a village in the center of Andar District, Ghazni Province, is safe, its bazaar, small hospital, and school intact. For now.

The operation ends and I'm given a choice. I can go home or I can stay in-country and take a desk job at the company command headquarters at Forward Operating Base (FOB) Ghazni, about eighty-five miles southwest of Kabul. The battalion of the 10th Mountain Division is here, retrograding, along with a brigade of Polish soldiers.

I accept the position, but after being a Green Beret field operator, working logistics behind a desk feels more like a penalty than a reward.

August 28, 2013, dawns clear and gorgeous. It's close to the end of my deployment and I dress in full gear for a "change of command" photo. I also bring along my coolest weapon: a .308-caliber SCAR MK20, a sniper support rifle (SSR) designed for both long-range and close-quarters combat.

Pilots get a bad rap for always making sure you know that they're pilots, but I joke that snipers are way worse. I go all out, so that anyone who sees this picture will know they're looking at a super cool sniper.

After the photo, I set my gear down and go grab coffee in the med shed with my medic buddy Scotty.

Suddenly, we're rocked by an explosion. It feels like a giant grabbed the building and shook it. I'm thrown to the floor, along with all the medical equipment. The dust inside is as thick as fog.

We've been hit directly by artillery, I think, getting to my feet.

Scotty and his patients are fine. I'm fine. Right now, I need to see where we've been hit and let the base know everyone in the med shed is okay.

Outside, there's so much dust hanging in the air, it's nearly obscuring the blue summer sky and amping up the confusion. Everybody is thinking *their* building got hit by artillery, but there aren't massive injuries, so there's no way we all got hit.

I hear small arms fire. It isn't ours. I'm hearing AK-47s and Soviet-made machine guns called PKMs and explosions from RPGs, and roughly half a mile away, near the rear of the base, I see a monstrous, fiery black mushroom cloud. The plume is massive, the biggest I've ever seen. Watching it, I feel small.

The explosion, I'll learn later, is caused by a nearly five-thousand-pound Vehicle Borne Improvised Explosive Device, or VBIED. Our base is under attack—but instead of coming through the main gate, the insurgents are coming through a sixty-foot hole in the rear perimeter wall.

Clearly, the enemy has a plan of some kind.

I've got to get out there.

My stuff is still all laid out where I left it, ready to go. I put on my kit and helmet and forgo the heavy gun belt. I conceal my Glock 19 pistol inside the waist of my uniform pants and grab my SCAR MK20 sniper rifle.

Now I need to find a way into the fight.

I spot a soldier driving one of our Toyota Tacoma pickup trucks. It's not a fully up-armored vehicle. More like half—just the doors and the back of the truck, not the glass. I bolt toward it as another soldier, Sergeant First Class Nate Abkemeier, comes running up.

"Are we doing this?" he asks me. "I'm driving."

We commandeer the truck. As we pull away, Nate almost runs over a four-wheeler. I recognize the driver. Sergeant First Class Andrew Busic is another Green Beret buddy of mine. Drew's about to drive into the fight, only his truck doesn't have a stitch of armor.

"Hey, Drew," I yell. "Get in with us. You'll get shot up in that four-wheeler."

He jumps in the back of the Toyota.

And I jump out.

The SF compound has its own gate. I need to protect the physical integrity of our compound, to prevent the Taliban from entering. I need to close the gate.

"Get in the truck!" Nate yells to me.

There's a three-story building overlooking our base. It's located 180 degrees from the blast — and the Taliban has occupied it. They were hiding in the building, waiting for the VBIED to go off. Now they're sprinting from cover, sixty, eighty, maybe as many as 150 guys. Some are shooting at Drew and Nate while I'm locking the gate.

"Get in the truck!"

As I make it back out through the foot gate, I hear a ton of fire coming from the breach in the perimeter wall.

I jump back in the truck, and as we drive away, another four-wheeler drives next to us. It's manned by Matt, a Green Beret medic, and Army Chief Warrant Officer 3 Mark Colbert. These are my guys. We're all working at headquarters.

"You guys going in there?" Matt asks, pointing.

"Yep," we reply. "Let's go."

We give each other a thumbs-up and speed off together.

The Taliban has a huge amount of weaponry. RPGs and recoilless rifles and a mortar section.

I'm sitting in the passenger's seat, my MK20 lying lengthwise on my lap.

"Slow down," I tell Nate, as we pull onto the airfield. "I'm going to dismount and climb that wall and employ my sniper rifle. I can hear these guys. They're on the airfield, about a hundred meters outside the camp."

As Nate slows, Matt's four-wheeler pulls in front of us. He edges toward the opening of the airfield.

They're immediately slammed by a monstrous amount of small arms fire. Rounds are hitting everything all around them. Both Matt and Chief Colbert get hit.

We've all rehearsed battle drills endlessly. We can't see where Matt and Chief Colbert are taking fire from, but we *can* see where the rounds are landing. Nate pulls our truck in between them and the incoming fire to create some cover. When Nate stops, I'm going to kick my door open and provide covering fire over the hood while Drew and Nate climb out after me to drag Matt and Chief Colbert to a better position.

But as Nate starts to turn, I see about a dozen Afghan Army guys standing in a semicircle about ten to fifteen meters away. The truck stops and I kick the door open, wondering what they're doing over here, facing the wrong way.

Then I get a closer look. These guys aren't Afghan soldiers — they're Taliban soldiers wearing Afghan uniforms.

They bring up their weapons and start shooting the shit out of our truck as they scramble for cover. It's then that I realize there's no protective armor inside the doors. The armor panels have been replaced by plywood.

My sniper rifle is a little over forty inches long. I jerk my rifle up and I spin. The motion accidentally shifts the truck into neutral. Nate doesn't realize. He's flooring the gas and not going anywhere.

I stick my arms out the door. As I present my rifle, the charging handle hits the doorframe.

I fire off one .308 round. It hits the dirt in front of the lead insurgent and throws rocks and gravel all over him. He ducks his head. I squeeze the trigger again.

My rifle jams.

Everything turns slo-mo.

My rifle has never jammed before—or since—but it's jammed now.

I see these guys staring at me. I see the muzzle blasts from their AK-47s. I'm not going to die sitting in a truck seat. I'm going to buy Nate and Drew some time.

When your rifle goes down, your pistol comes out. My Glock is drawn before my feet hit the ground. I shut the door behind me, trying to keep any incoming enemy fire from hitting Nate.

I'm an exceptional pistol shot by any standard, but I've never fired one in combat. Now I'm going to engage Taliban fighters using only a Glock 19 handgun with ball ammo. I target the nearest group of fighters.

There're three of them. I hit the lead guy in the pelvic girdle and he instantly collapses. I'd always heard stories about how a nine mil doesn't have any knockdown power, that you need a .45 to create any effect. I thought I'd have to put fifteen rounds in him.

His fellow soldiers don't run to help. They run away.

Clear the front site post. If you don't, you're not going to hit anything — and you'll die out here.

The best cover is your own muzzle. That's always been one of my mantras. As I squeeze off rounds and advance, Nate manages to jam the truck into reverse. He floors the gas, yanks the wheel and spins around, and ends up crashing into a wall behind me.

I keep firing, closing the distance between me and the enemy. Keep waiting for a bullet to hit me and for it to be over. I just want to make as much of an impact as I can before that moment comes.

The bullets keep missing me, but they keep hitting Nate and the Chief. I see the two of them hit the ground like sacks. Drew is trapped in the back of the seat. He's shaking the doors, trying to get out to join me, and he's catching hell from all the incoming fire.

I end up driving the enemy back and away from us.

My pistol is now mostly empty. My rifle is jammed. I'm alone with at least a dozen or more insurgents somewhere in front of me. The closest one is the guy I hit in the pelvic girdle. He's not dead, but he can't walk. He's lying on the ground, and he's firing his rifle. At me.

I've got to find cover so I can reload my weapon.

I duck behind a little water tank. Then I remember I have a hand grenade.

I can use it to create just enough time for me to fix the malfunction in my rifle.

I grab the grenade, pull the pin. I lean out and gently toss it in the direction of the insurgent I wounded.

He starts firing at me, then the tank. He doesn't let up. Little pieces of white plastic rain down on me as I'm locking the bolt of my rifle to the rear and ripping the magazine out. I'm digging out the mess in the chamber when the grenade detonates.

In the movies, if a guy gets hit by an explosion, he gets blown straight up in the air because that's cinematic — but in real life, usually the person just folds over. But this explosion actually *does* blow the insurgent straight into the air. His arms and legs are pinwheeling, his body flopping around, and I'm thinking, *That's not a thing you normally see. Nobody is ever going to believe that that happened.*

I've got my rifle up now. I'm looking around, but I can't find anybody. The insurgents who were here moments ago seem to have retreated.

They know something I don't. What are they planning to —?

Fire to my rear — the distinctive *snap-thump* sound made by a rifle.

A round cracks, then hits the wall eight inches above my head. I look over and see a guy lying on the airfield, in a sling-supported prone position at a hundred meters out. I know the exact distance because he's lying at the edge of the area where we do our sprints.

He's taking well-aimed slow fire at me. He's missing, but not by much. If he aims for a body shot, it'll probably be a different story — but he's going for a head shot.

I drop to a knee, look through my sniper scope and hold the notch where the guy's throat and clavicle meet up.

I hold center.

Pull the trigger.

He's gone. Vaporized off the planet.

I'm startled by the tremendous thunderclap of sound that follows. I start looking around, thinking the Polish Armor has shown up with a tank, maybe hit him with a main gun.

That's when I figure it out: The guy was wearing a suicide vest. I must have shot the vest and detonated it—which, I'm guessing, also explains what happened moments ago with the other guy, the one I saw sailing through the air like in a scene from a movie. My grenade must've detonated *his* suicide vest.

That's why his insurgent buddies retreated. They didn't want to get blown up. Maybe these guys have some sort of pact where they detonate their vests if they go down, and they suspected that's what he'd do.

Which brings up a disturbing question.

How many of these fighters are wearing suicide vests?

My guys are yelling for a medic, for help, for support. They're behind me, somewhere around the corner. I can hear them. Chief's hit. Matt's hit. Everybody back there's hit. But if I run back to them, the fighters will be right on top of me. That's not going to help my guys.

I scan the area, looking for more insurgents, wondering what their big plan is. I'm worried they're making their way back to the camp. I've got to delay their movement. I decide to go close with the enemy, figuring if they're running away from me, at least it'll keep them off my guys. Got to buy time for my guys until support arrives.

When the insurgents scattered, I saw a few of them run

down this little lane in front of me. I get about halfway down when I engage three or five fighters, but it feels like a hundred guys because all I see are muzzle flashes. They're firing from cover, about twelve to fifteen meters away. I'm too close to use my sniper scope, but my rifle has a little .45 optic on the side that allows me to line my sights up.

I start playing Whac-A-Mole. I focus on one bad guy while his buddies fire at me. I move toward them, eyeing a generator panel ahead. I need to get there for cover because I'm running out of ammo.

I don't make it.

I dump the mag. Now to reload. As I pull my muzzle up, the nearest fighter breaks from cover. He throws his rifle into his sling and screams, *"Allahu Akbar!"* and starts sprinting toward me.

He's wearing a suicide vest.

I'm faster at reloading than he is at sprinting. I drop my muzzle down as I send the bolt home. I start firing and move behind the generator panel. He's seven, maybe ten meters away when my third shot detonates his vest.

The generator panel absorbs the fragments from the explosion, but I still get rocked down to the ground. I'm not knocked unconscious, but it knocks me down, rings my bell. I'm TKO. Confused as to what's going on and where I am.

Another fighter breaks cover. He looks over his rifle as he walks toward me, calmly cranking off round after round. He's trying to shoot me in the face as I'm lying there, yet all his shots keep landing short. He's making eye contact with me but missing, because he's looking at me instead of looking down his weapon sights.

I jerk my rifle up and start hammering away at him. He collapses in a heap.

Now I've got to engage the other fighters to his rear.

We're exchanging fire when a five-hundred-gallon tank full of aviation fuel detonates into a huge fireball. It's intensely hot. There's black smoke everywhere, tons of it. I use the opportunity to move around the corner and reload again.

I'm gagging on the thick, sooty smoke when Drew appears.

A round hit must have hit the safety locks on the truck earlier, trapping him in the back seat. He's covered in huge cuts from where the incoming fire hit the truck—the bullets fragmented, deforming into big, fat ninja throwing stars and bouncing around all over the place.

"I know where they're at," I tell him. "Let's go get them."

Together, we turn toward the lane. We're doing cross coverage and getting closer to where the body of the last bad guy I shot is now smoldering and smoking. Drew is about to step over him when I yell, "Stay away from the bodies. They're all wearing suicide vests."

Then, as if on cue, the guy's vest goes into what we call low order. It doesn't detonate, but it starts burning like a gigantic blow torch—an intense plume that shoots twenty or thirty feet in the air. It's hot and nasty, and I'm so close that the heat feels like it's cooking my skin. We duck behind another generator panel to wait it out.

Aviation fuel smolders behind us. The area is thick with smoke, and every now and then the wind stirs up and scatters the smoke just enough for us to see a bad guy or two. We engage them, only these guys are different than the others.

These fighters are carrying a tremendous amount of ammunition. They have under-barrel launchers for their AKs, and they're all wearing belts carrying about twenty hand grenades. One guy shoots at us while the other throws grenades as fast as he can, over and over and over again.

There's a constant *thump-crump, thump-crump* as grenades detonate. I'm trying to line up my sights, but fragments from the explosions are blowing rocks and dirt in the air, and although the junction panel absorbs the force, the concussion is like a jab to the face every time. I feel like my bones are getting split in half.

Drew and I are constantly getting concussed, but whenever there's a clearing and we see a guy, we take a couple of shots. I'm working my cover when something whacks the top of my plate carrier, near the base of my throat. It sounds like a loud crack.

I look down. A hand grenade is trapped between my admin pouch and the junction panel.

If I step back, the grenade will fall between my feet. I'm trying to keep it pinned against the panel so I have control of it, but I'm also trying to slap it away from me like a poisonous spider. So I start pawing at the grenade and eventually manage to rip it away from me. There's another detonation. Drew and I get whomped again.

I'm clawing my way back up when another grenade hits the back of my knee. Drew and I kick it away furiously. The grenade explodes, blowing us down again.

"We've got to get out of here," Drew says. "They're going to kill us."

He yanks me up. We only take three or four steps before getting blown down again. We fall together to the ground in a tangled heap. This time I land on top of him. I've got forty pounds on Drew, and I'm crushing him. It's all a big mess.

I'm trying to get my rifle up when I look down and see a severed forearm on my rifle. The explosion threw this severed arm so hard it damaged the butt stock of my rifle. I stagger to my feet and try to fix the butt stock. Drew drags me down an alley. We make our way back to the corner, back to where we were first ambushed.

Then I see Chief Colbert come limping up. I thought he was dead — was *sure* he was dead. But here he is, not only standing on his own two feet, but not even looking like he's hurt too bad.

I want to run over and hug him, I'm so elated.

He looks us over. Grins.

"What are you boys doing?"

Nobody is shooting at us, but we're not in a safe position.

Drew says, "We know right where they're at."

"All right," Chief Colbert says. "Let's go get 'em."

My pistol's empty. My rifle's empty. I start fishing around in my kit for a fresh magazine. There isn't one. All my magazine pouches are empty.

"Chief," I say. "I'm out of ammo. You have to take point."

"What the hell are you talking about?"

"The only magazine I have is the one in my rifle. I've got two rounds left. I can't take point. You have to do it. But hey, don't worry, I'll cover you."

He shoots me a look that says, *You're going to cover me with two bullets. Thanks, asshole.*

Someone calls out behind me. I turn around and see a Polish Army lieutenant, Karol Cierpica, and Staff Sergeant Mike Ollis, a soldier from the 10th Mountain Division, sprinting toward us.

"We want to come with you," Ollis says.

He isn't wearing any body armor, just his combat top. He doesn't have any gear on his rifle, just a single magazine. I'm about to say something about how he shouldn't be over here with no gear and no ammo when I look down at my own rifle, loaded with only two shots.

"If you want to go," I say, "let's go."

They stack up behind us.

Another guy joins us. Lieutenant Turnipseed, a Navy SEAL. We have a good formation. I know we're going to dominate the enemy. There's just so much energy radiating off these guys, they're ready to hunt. Win.

There's no other place I'd rather be than right here. We've been taking it on the chin, but now we're coming back. I let the moment sink in as we march together to the gates of hell.

But the gates of hell are quiet now. There are bodies everywhere—but not whole bodies, just pieces. Arms and legs and rib cages. Spinal cords and organ meat. Dark smears of blood and bile on concrete.

The indirect fire has slowed down because the Polish armor has moved up and filled the breach. They've pushed the insurgents off the camp, and now they're hammering away at what's left of the Taliban.

We're standing there, watching, when we see movement from behind a nearby pile of bodies. It's one of the fighters. He sits up and throws two hand grenades as he screams "Allahu Akbar!" and yanks on his vest, detonating it.

Everyone scrambles for cover as the grenades bounce across the ground. I'm backed by the junction panel again, wearily thinking, *Oh, man, these again.*

I turn my head away from the explosion as the grenades detonate. Man does it hurt. My bones feel as though they've been pulverized into dust. By this point, I'm thinking, *I'm a pro at getting grenaded.*

I hear gunfire behind me. When I turn around, I see Drew yelling something from the other side of these clear shipping containers—and I see him shooting at our rear.

A Taliban fighter is trying to attack us. Drew is trying to fight him off. Mike Ollis is also nearby. He's rendering first aid to Polish Army lieutenant Cierpica. Mike moves himself between the attacker and Cierpica and starts firing.

I raise my rifle and fire my last two rounds.

I don't know if I hit him, or if it was Mike or Drew or maybe all three of us, but the fighter's vest detonates. Mike takes the brunt of the blast. It launches him backward ten feet. He hits the ground right in front of me and rolls and comes to a dead stop. He isn't moving.

My rifle is empty. I drop it. The only weapon I have left is my knife. I take it out, thinking, *I'm going to run over to the fighter and cut his throat*. Then I realize that's ridiculous since the insurgents are wearing suicide vests.

There's an AK on the ground. I grab that instead and I run back to Mike. His eyes are open. He looks startled, but when he sees that it's me, he relaxes a bit. His left arm is partially amputated.

I pick up my rifle, sling it over my shoulder, and try to drag him by the shirt. It keeps stretching off him, so I grab him by the buckle of his belt and carry him back to the compound. I have no idea if it's safe, but it's got to be safer than where we're at.

Mike's arm is bleeding badly. I drop the AK and put him down. I talk to him, and as I get a tourniquet on his arm, I see he's also got a leg injury. It's not bleeding yet, but I don't want to take any chances. After a blast injury, when someone starts to relax a bit, things can turn bad. Fast.

He's no longer conscious. I slap his face, urging him to stay with me, but he's not responsive. As I throw a tourniquet on his leg, I look around for my guys. They appear to have run off. I have no idea where they went. I'm alone, and I'm worried about someone walking up behind us.

Mike is fighting to breathe. I rip up his shirt. He's got a huge depression on his chest.

I'm at the camp. Literally. The hospital is a minute away. I know Mike's best chance is there, not here with me trying to figure out what's going on.

I spot a civilian with a vehicle, and load Mike up in it. As another soldier comes running over, I say, "I'll cover you two. Get this kid to the hospital."

Then I turn back.

I'm alone, with absolutely no ammunition, and I need a gun.

I return to the spot where I dropped the AK earlier. Something's rattling inside. I take a closer look. The rifle has bullet holes through the bolt and operating system. *Crap.*

Fortunately, rifles are scattered everywhere. I pick up another AK. This one has rounds through the magazine right next to the receiver, which is now stretched apart. It doesn't work, but it's got a good bolt. I rip the bolt out and throw it back in the other rifle. Now I've got a working weapon.

I'm scurrying around, grabbing grenades for the launcher and rifle magazines off the ground when I see my new incoming commander, Major Kaster. I tell him what's going on, and then the two of us start clearing all the buildings in the area.

People have trickled back into the compound when we return to company headquarters. Everyone looks shot to pieces. Drew sees me, wants to know how badly I'm wounded.

"I'm not hit."

He looks at me like I'm an idiot. Every single part of me is covered in blood. I look like I've been dipped in it. Every time a suicide vest detonated, it threw aerosol sprays and splashes of blood.

"You're shell-shocked," he says, and starts sweeping me over for wounds. The only injury I have is a piece of unfired brass from a rifle stuck in my arm. I pull it out myself and go to check on Mike.

The bleeding in his leg, I'm told, is from a severed femoral artery. He's also been shot in the stomach. When the doctors go to do chest compressions, they discover that all of his ribs are broken. The blast from the suicide vest inflicted extensive internal injuries.

He doesn't make it. He sacrificed his life to save Polish Army lieutenant Cierpica, who lay on the gurney next to Mike the whole time. Cierpica is able to tell Mike's grieving father that the doctors did everything they could to save his son.

Chief Colbert and Drew receive the Silver Star for their valor in battle. Nate receives a Bronze Star with V device, one level below the Silver Star. Mike is awarded the Silver Star posthumously — and becomes a hero in Poland, where he is also posthumously awarded the Army Gold Medal, the government's top award for a foreign soldier.

Lieutenant Cierpica and his wife later name their newborn son Michael.

When I first tell my story, I underplay it a bit because it's so over the top. So crazy.

An investigation is launched. Investigators come to the base. A rarity. Step-by-step, I take the investigators through the various crime scenes by retracing my footsteps. I tell them, "I was standing here when I threw my grenade." Then as I look down at the gravel, I find my grenade pin.

"This is bonkers," the investigators keep telling me. They find all my brass.

Since it happened on base, both witnesses and video corroborate different parts of my story. Investigators count three hundred–odd rounds in the truck and in the wall where the gunfight happened. The shots were all fired inside of twenty meters, and not once did I get hit.

Which drives Drew crazy because he was getting pummeled by rounds, shrapnel, and grenade fragments.

When I think about the attack, I always go back to that moment when we got organized and decided to engage the enemy as one. The way we assembled into a synched stack and moved aggressively, right into the chaos. To be with those guys, at that time, on that day, is probably the proudest moment of my career.

It's the epitome of soldierly virtue on the battlefield.

Alan Mack

2nd Battalion 160th SOAR Flight Lead and Standardization Instructor Pilot, US Army
Conflict/Era: War on Terrorism (Afghanistan)
Date of Action: March 3, 2002
Distinguished Flying Cross

"There's always room for one more Ranger."

"Not in Afghanistan," I say. I'm speaking to my superior officer, the Fifth Group commander.

"You're going to take our teams in and link up with a warlord," he's telling me and the other flight lead. "They're going to do what's called a UW, or an unconventional warfare campaign. We'll use them to take the Taliban down."

I choose my next words carefully. "How much a team weighs determines how far we can take them. The math doesn't work out to carry even half the team in over these mountains."

It's late 2001, the early days of the War in Afghanistan. Defense secretary Donald Rumsfeld calls the planning area and says, "You get those teams in now, period," and hangs up. Then the team leader comes to me and says I need to take the team in my Chinook. Not half the team. The *entire* team.

"I can't," I tell him. "It's a weight issue."

"Figure it out. We leave tonight."

I lay out all the different ways the operation can go sideways.

"Look," the team leader says. "Just get me there."

It's a common refrain. Throughout my career, pilots are viewed as nothing more than glorified taxi drivers — fly me from point A to point B. Our job is way more complicated than that.

The Chinook cargo helicopter is one of the fastest helicopters in the world, capable of speeds of up to two hundred mph. With an airframe the size of a Greyhound bus topped by a twin-engine, tandem-rotor system, the joke on the Chinook is that it looks like a dumpster stuck between two palm trees.

The only way I can drop weight to an acceptable level is to use less fuel, which means I'm going to have to refuel in the air. This mission's lowest altitude is nine thousand feet. I'll have to climb to fourteen thousand feet to refuel. Today, completing an air refuel is the norm. During this time, refueling at over six thousand feet brings trouble.

ODA 555 has taken three of the four Chinook helicopters. I have the fourth. My wingmen will be a pair of armed Blackhawks, what we call DAPS — direct action penetrators. The Blackhawks will fly into the mountains as far as they can and then wait for me while I'm out of comms for three hours.

The first mission objective is to meet with warlord Fahim Khan, the new leader of Afghanistan's National Resistance Front (NRF), a group comprised mainly of former Afghan National Security Forces (ANSF) who are committed to fighting the Taliban.

Khan took over leadership when Ahmad Shah Massoud was assassinated. Khan doesn't like to meet people — he's afraid of getting assassinated himself — and while he's secretive and not well liked, he is providing us tactical information, which has, so far, been sparse and contradictory.

Special Forces Operational Detachment Alphas, or ODAs, take the lead on the meeting with Khan. These elite, highly trained teams specialize in advanced weapons, language, demolitions, combat medicine, and advanced combat tactics.

The first flight takes off carrying ODA 555. I launch thirty minutes later in the other Chinook. Then it's time to refuel.

I always make fun of movies where they show the helicopter pilot moving the flight controls as though he's churning butter when, really, it's all in the wrist.

The handling characteristics are different at higher altitudes. The controls are mushy and don't respond as quickly. I steady my forearm on my thigh because I can't be fast or aggressive. *Slow is smooth, and smooth is fast,* I keep telling myself as I climb.

We manage to refuel without incident. I descend and then fly across the Amu Darya, the river between Uzbekistan and Afghanistan, and encounter a sandstorm that's several thousand feet thick.

"Hell," my copilot, Jethro, says, "I can't see anything."

I look out the side window, the one down by my leg, and see we're flying along a wall of rock.

My battalion commander, sitting in the jump seat behind me and Jethro, says, "Al, what do you think?"

"Sir, I think we just TF." Terrain Following Radar will allow us to hug low spots in mountain passes and valleys even when it's dark or when, like now, we're caught in bad weather with zero visibility.

I'm waiting for the battalion commander to say, "Turn around." Instead, he says, "Okay, execute."

The DAPs don't have this equipment. They suck up as tight as they can and use their night-vision goggles to track the fluctuations of my engine exhaust. Expanding engine glow indicates a climb; dimming indicates descent.

They last five minutes before they abort and head back. Now it's just my aircraft. We have no communication capabilities or ISR—Intelligence, Surveillance, and Reconnaissance. We're completely alone.

"Jethro," I say, "climb like your life depends on it, because it does."

We're climbing and I'm cycling through digital maps when I see it—a twelve-thousand-foot mountain. We don't have the necessary power to get over it. I turn the aircraft 20 degrees to my right and descend.

We get past the mountain. As I rejoin our original course, everyone is quiet. Tense. They're all thinking, *Oh, my God, we almost died.*

We've all seen a movie called *O Brother, Where Art Thou?* I quote a line from the George Clooney character. "Damn, we're in a tight spot."

"Yeah," Jethro says. "Damn, we're in a tight spot."

The crew chief starts laughing. The tension evaporates.

I have one last hurdle. To land, I have to break out of the

clouds and then descend five thousand feet while avoiding taking on fire from a ZPU 23-4, a Soviet-manufactured light tank equipped with four water-cooled 23 mm auto cannons. I need to keep a little hill between us and the gun or, with a rate of fire of 850 to 1,000 rpm, it can turn us into Swiss cheese.

I execute a series of S turns, where we lose altitude laterally instead of flying straight ahead. We land on the ground, the dust so fine it might as well be talcum powder. When the dust settles, we're surrounded by Afghan men with mujahideen scarves covering their faces.

An Agency guy is supposed to be waiting here, but I don't see him—or a single American. I feel a stab of panic, wondering if we've just landed in the middle of an ambush.

The team gets off the helicopter and starts hugging the Afghans. I get the thumbs-up to return to base. When I arrive, me and the other flight lead sneak off to a dark corner and share a bourbon.

"What do you think?" he asks.

"I don't think we're going to survive the next flight."

Only we do. We survive the flight and the next one and the next even though we encounter new and unforeseen problems. The only time we don't fly is if the weather doesn't permit it. We can't ever say no.

In early 2002, Taliban and Al-Qaeda forces gather in the Shahikot Valley of eastern Afghanistan. Bin Laden and his right-hand man, Ayman al-Zawahiri, who is believed to be the operational mastermind behind 9/11, are both somewhere in the valley.

Operation Anaconda is scheduled to kick off in February.

All non-governmental organizations like the Red Cross are asked to leave the area. Two of my seven Chinooks, Razor 03 and Razor 04, are assigned as the high-value target takedown team. If Bin Laden and/or Zawahiri show up, I'll carry SEALs to go get the top two most-wanted terrorists in the United States.

The other Chinooks will be carrying Special Forces coalition teams. They'll be placed on various mountaintops overlooking high-speed avenues of approach or departure. If Bin Laden and/or Zawahiri try to flee, these SF teams will call for fire to stop them or seal off the area.

Bad weather delays Operation Anaconda to March 1. Word gets out, giving the enemy time to prepare.

There are only so many places a helicopter can land in the valley, and they're the exact same areas the Soviets used in their war with Afghanistan. The enemy knows this, and their previous fighting experience with the Soviets has equipped them with range cards—approximate coordinates to launch successful mortar attacks.

When the 101st and 82nd go in with their Chinooks, they get their asses handed to them, through no fault of their own. The overwatch teams situated on the mountaintops are designed for avenues of approach. The Chinooks are in a blind position.

US forces consider pulling out on the first day. They decide to stay and fight and manage to gain a foothold. Now we need eyes on a piece of key terrain to help us see the battleground and coordinate.

I'm ordered to take my SEALs to Takur Ghar. I'll drop them at the base of the mountain. From there, they'll make their way to the top, which will allow them to observe the series of cave complexes the enemy is using.

The bad guys are concentrated in what's called the "white whale," a terrain feature that from the air, looks like a whale on the water. The problem is, every time I take off, I'm ordered to turn around because the 101st is calling in B-52 strikes on the area. It's a ten-minute flight back to Forward Operating Base Gardez, but these multiple takeoffs and returns have us burning through gas—and there's no additional gas or fuel at the FOB. I need a certain amount of fuel in the tanks to complete the mission and get us back home.

I shut the aircraft down to conserve fuel. When we get the go-ahead to proceed, I start the helicopter back up and discover that one of the engines has a maintenance problem. We don't have what's called single-engine capability. If one engine fails, the other doesn't have enough power to continue flying, resulting in a crash.

I call Bagram Air Base. Razor 01 and Razor 02, I'm told, are just returning from a special operations run. Both aircraft will be refueled and flown to me so we can complete our mission.

But it's a forty-five minute flight. I've been turned around so many times that sunrise is coming, so there's no way I'm going to be able to place the SEALs at the base of Takur Ghar and have them walk to the top under cover of darkness.

Which is why I'm ordered to take them directly to the *top* of the snow-covered mountain.

"I haven't seen any imagery," I say. "I don't even know if there's a landing area up there."

One of the SEALs says, "I've seen imagery at the top. There's a place to land, you'll be good."

I take him at his word. If there was a problem, he would tell me. Plus, we're going to send an AC 130 overhead, check out the landing zone, and report back with any issues.

Razor 04, my other helicopter for this mission, doesn't have enough fuel, so we end up swapping out both of my helicopters for Razor 01 and Razor 02.

The SEALs use tactical headsets that are incompatible with the aircraft, so we've installed a pane of plexiglass where we can use a grease pencil to pass along messages. This saves the team leader from having to put on an aircraft headset that allows him to talk to me but not to his guys.

For seven months, I have so much ordnance fired at me while in the air that I get numb to it. I'm expecting to receive some fire as we approach the LZ, but it's quiet. We land on top of the mountain, at a 12,500 elevation, without a problem.

The team leader takes off his aircraft headset. He's starting to put on his tactical headset when, just off to my right, I notice a DShK mounted on a tripod. The Soviet-made heavy machine gun fires 14.5mm anti-aircraft rounds.

"Get the team leader back on comms," I say.

He takes his helmet off, and when he puts the headset back on, I say, "Team leader, you've got a DShK, unmanned, at two o'clock, fifty meters."

There's also a donkey tied to a tree. Obviously, somebody's up there.

We've been on the ground for maybe fifty seconds. The team

leader puts his tactical headset back on and then he and his men start moving to the ramp that will allow them access to the mountain.

"Sir," my left gunner says, "there's a man at about nine o'clock. He just popped up from a berm."

"Is he armed?"

"I don't know. He just disappeared."

The day before, the crew of an Air Force AC-130 gunship mistakenly fired on friendly forces, killing an American SF soldier. That fratricide changed the rules of engagement. Essentially, I have to get shot at before I can return fire. That's a hostile act.

Hostile intent is more difficult to determine. Any friendly would know to stay down.

"If he pops up again," I tell the gunner, "kill him."

The guy pops back up, this time at an eleven o'clock position. This time, he fires a Rocket Propelled Grenade.

It's like a scene from the movie *Black Hawk Down*—a lava lamp flying at me with sparks shooting out of the back in slow motion. The RPG hits the aircraft just behind my seat, in front of the fuel tanks. Another foot to the aft and we all would have blown up. Another couple of feet to the front and I would be dead.

There's an explosion inside the aircraft. The soundproofing catches on fire as the RPG rips through the left ammo can for the helicopter's miniguns and then exits next to the cab door, missing the two crew members in the front and the two gunners and the team leader.

It's a miracle no one is killed.

The crew chief in the back yells, *"Fire in the cabin, fire in the cabin."* I can hear his voice over my headset, but all the cabin displays are blank. The RPG hit the two main electrical systems plus the auxiliary, so now we're running strictly on battery.

And now bullets are hitting the aircraft.

The miniguns aren't operational. They run on AC power, and with the generator out of commission, we have no way to defend ourselves.

Are the SEALs inside the aircraft? Do they want to get off? Are half of them outside? I just don't know. My front guys aren't talking to me because they're in shock, dazed from the explosion.

I pull power to take off. I hear the rotor blades droop because the engines are in backup mode. Judging from what I'm hearing, I think I've lost an engine. If so, there's no way I can fly off, which leaves me with only one alternative.

I'm going to have to dive down the mountain.

My goal is to gain air speed, which will help me gain rotor RPM in the descent, so I can essentially, with one engine, have a controlled crash at the bottom of the hill.

The DShK is shooting at us. I execute the flight maneuver, diving at 130 mph and trying to use the trees for cover. What I don't know is that Navy SEAL Neil Roberts has fallen out of the aircraft, into twelve feet of snow. The left rear gunner, a guy named John, tried to grab Roberts and was pulled out with him. Now John, secured in a canvas tether harness, is hanging from the bottom of the aircraft, his feet tickling the pine trees as I'm diving down the hill.

The crew chiefs up front have finally come out of their daze. One says, "Sir, the other engine is running. You can level off."

"Are you sure?" The displays are dead. We only have sound to go on.

"Yes, sir, I'm sure. We're good."

I level off, and sure enough, I can maintain flight. About that time, Pete, the other gunner in the back, pulls John up hand over hand. Then he gets on an M60 machine gun and returns fire.

I have a new problem. I can't move the flight controls anymore. We've lost hydraulic fluid, and we've taken a few rounds on the transmission. Any second now, we're going to come apart.

"I'm sorry, guys, we're done," I say. "I can't move the controls."

The crew chief in the back has cans of hydraulic fluid. He uses a screwdriver to poke holes in them and then pours the fluid into a reservoir meant for in-flight servicing. There's a tiny handle shaped like a T that he needs to pump. He's back there, working like a maniac, when I'm told we lost a guy in the landing zone.

I ask for a head count. Suddenly, the controls return.

Everyone is present except for Roberts. I have control again and turn back to the landing zone. We're a mile or two away, on a descent to land on top of the hill, when the controls lock up again.

The crew chief adds another quart of hydraulic oil. The controls come back online. I realize I've got about fifty seconds to do something.

If I land, we're not getting back off the ground. I'm carrying SEALs, plus my crew. I'm responsible for them.

"Razor Zero Four is still out there," I say. "Have him pick up Roberts. Let's see if we can get on the ground safely."

I make a turn back to the left and can see the entire battle unfolding below me. If we have a three-hundred-foot rate of descent, at seventy mph, we might survive—provided the controls don't lock up again.

My copilot starts calling out air speed and altitude.

"Where are you getting that information?" I ask.

He taps a finger against the backup altitude indicator. It operates on battery power—that's how he knows our speed and rate of descent. Up until this point, I've been suffering from tunnel vision. Now my vision opens up more so I can start assessing our situation a bit better.

We descend to the bottom. We're about thirty feet off the ground when the controls freeze up again. The aircraft is in a landing altitude, but we're sliding to the right, heading to a hill that's higher than me. If I can't stop the slide, we're going to hit the hill sideways, roll over, and explode in a grand fashion.

There's a saying in aviation: Never quit flying the aircraft. I can't move the cyclic, which is the control stick, so I try the pedals, which make the aircraft rotate at a hover. I push hard on the pedals.

It aligns with the landing direction. Now we're going to hit the hill straight on instead of sideways.

As we come in, the landing gear starts brushing the ground. I push on the power lever, which, in a Chinook, is called a thrust. We settle on a cross slope of about 15 to 20 degrees, meaning the left side of the aircraft is much lower than the right side. Flying this way, at this angle—it's not a good place to be.

We manage to land—and survive.

I use the rotor brake, which stops the rotor in about one revolution, and shut off the aircraft. We get outside, a bunch of pilots and crew chiefs who are now part of a SEAL reconnaissance team.

A pilot on the ground—that's not the natural order of things, and I'm a little nervous. I'm good with weapons, but pilots are not ground-level tactical people. No one is firing at us yet, so we set up a little perimeter. Someone brings out the M60.

I go around checking on people. By the time I reach the team leader, I'm out of breath.

"Dude," he says. "You need to breathe."

"What?"

"Breathe normally. Just breathe normally." Then he points up at the sky and says, "We've got an AC-130 overhead. We're talking to the Air Force operator, the Omega elements. And Razor Zero Four knows about us now, so they'll come and get us."

He's right about the AC-130. It's firing down at the enemy, providing us cover. I'm plotting our position when the SEALs say they want to walk up to Takur Ghar.

"It's ten miles away," I tell them.

"No, it's right here. We're near the base of the mountain."

"That's not Takur Ghar." I show them the map. Sure enough, I'm right.

Now the discussion turns to blowing up the aircraft. We can't allow it—or the technology on board—to fall into enemy hands.

We decide to wait. It can be done with bombs from air. Razor 04 drops us at FOB Gardez then returns the SEALs to the top of the mountain to rescue their fellow SEAL, Neil Roberts.

I make my way to the main operations room and call my brief in to Bagram Air Base. I'm told to hang tight at the lightly guarded base. Then a communications analyst intercepts key information that the enemy prepared to attack sometime before noon.

The soldiers stationed at Gardez are working with the British. As I get to work helping to secure the base, Razor 04 returns. The helicopter has been shot up by the DShK gunner. The SEALs, though, are on the mountain. When the gunner stopped to reload, Razor 04 dropped the SEALs off and then dove down the mountainside to avoid incoming enemy fire.

There's good news. Razors 01 and 02 are carrying a Quick Reaction Force (QRF) composed of Rangers. They're minutes away from Takur Ghar.

Razor 01 attempts to land where I was ambushed. The helicopter gets shot down from an onslaught of machine gun and RPG fire. As they fight for their lives against overwhelming odds, Razor 02 drops the Rangers off on a lower part of the mountain. Right now, they're fighting and clawing their way up the mountain.

It's time to prepare for a nighttime rescue.

I won't be going. A colonel from Bagram calls and tells me to get in "PZ posture"— shorthand for get to the pickup zone and be ready for the helicopters. It's dark and my Afghan driver is forced to maneuver the truck across a known minefield, so the load-in is already underway when I arrive at the Chinook.

I walk up the ramp and look into the cabin. Bodies are carefully arranged, but there's no escaping the smell of blood. Death.

The crew chief is yelling at me to get on board now. He's running out of gas. I take a seat on the ramp hinge. My left thigh brushes against the right thigh of Navy SEAL Neil Roberts. The luminescent dials of his watch are moving, which means his hand is moving, which means he's *alive*.

Then I realize my emotional state, my stress and fatigue, has caused my imagination to play tricks on me. His hand is moving from the helicopter's vibrations.

He's gone. Later, I learn that Neil Roberts, a member of SEAL Team Six, is the first casualty of Operation Anaconda. He didn't die from the fall. He died on the battlefield.

When people see all the bling and fruit salad on my uniform, they go, wow, look at all those awards. My first Distinguished Flying Cross — I'm very proud of it, received the award for essentially my first six months in Afghanistan where I pushed my aircraft to its limits, getting guys on the ground. A second Distinguished Flying Cross came my way after me and several other people made some poor choices and then happened to overcome a situation.

That's the thing about awards. I tell people I won them because either someone fucked up or I screwed up and survived. Or someone else screwed up and I fixed it.

Gary Wetzel

Private First Class, US Army
Conflict/Era: Vietnam War
Action Date: January 8, 1968
Medal of Honor

Tomorrow is my first flight in Vietnam. I stay up all night partying with the guys, getting to know them, and before long the sun is coming up and I'm in a helicopter armed with M60 machine guns, taking off to our destination.

I'm the oldest brother of nine siblings. I dropped out of school in the tenth grade. My parents served in World War II — my dad fought in the Philippines and my mother was in the Women's Army Corps (WAC) — and they both said to me, "Why don't you join the Army? Maybe that'll make a man out of you." I took their advice and enlisted on February 15, 1968.

My duty assignment is combat engineer. During my downtime at Fort Leonard Wood in Missouri, I'd go over to the range, where the guys taught me how to take apart the M60 machine gun. I always loved flying, and I found out that if I reenlisted for three years, I would get my choice of duty stations—which I did.

And now here I am, a redheaded nineteen-year-old kid from South Milwaukee working as a door gunner, sitting in my own little cubby hole in a Bell UH-1 "Huey" helicopter. We're a crew of four, all young guys, eighteen to twenty-one. The aircraft commander, the crew chief (and second gunner), and "Peter Pilot" (the copilot flying in the right seat), are all depending on me to make sure the guns fire.

Twenty minutes into the trip, I realize I need a toilet. When I'm confident nobody's looking, I lean out and take a leak with the force of a firehouse. The back rotor wash blows the waste all over me.

When we land, one of my buddies sniffs at me and says, "You pissed out of the chopper, didn't ya?"

"No, I didn't."

"Kid, we've all done it. Next time bring a canister with you."

I bring an empty coffee can with me every time I fly.

I fly every day. On average, a chopper goes down once a week. Half the time, the crew makes it out alive. When they don't—it hurts. I teach myself not to get too friendly or emotionally attached with anyone, because when the connection is broken it screws with my head.

You're here to do a job, I tell myself every time I board a chopper, every time I go up in the air. *You're a soldier, do your job.*

Door gunners have the shortest life expectancy. I fight hard to keep the fear at bay, or it will consume me.

One time, our helicopter is heading into a hot landing zone, and both of us door gunners are laying down suppressing fire. As the troops prepare to exit the aircraft, some of them speak to me. One takes the cross from around his neck, places it in my hands, and says, "You need this more than I do." Another gives me a quarter and a nickel and says, "Thanks for the ride."

On June 27, 1968, I'm flying out of a base camp with seven grunts. One soldier has brought on board a puppy that's about eight weeks old. When we reach about fifteen hundred feet, the whole chopper starts shaking.

I'm cooler than a son of a bitch. I look out, see that the tail rotor is shut off—and that's when we start doing what's referred to as 360s. The cabin of the chopper spins into the same rotation as the blades.

The pilot pushes in the left pedal and we auto-rotate down to the ground, looking for a spot to land. I start making a mental checklist. First: get the pilot out of his seat.

The pilot pulls up pitch and sets the chopper down hard in a sandy area. As the helicopter blade hits the ground, I jump out and rip off the pilot's door. I get him out of the chopper then discover—as the puppy jumps up and down around us—that one of the grunts has his leg pinned underneath. Using a machete as a shovel, I release the grunt, then grab the ammo stored underneath my seat and give it to the pilot. Then I grab the M60 and we book it to a nearby house. There, I bust off eight caps to keep Charlie's head pinned down.

Next thing I know, the enemy has *us* pinned down.

During the fighting, my skin starts burning. I look down

and see that I've set up the M60 on an anthill swarming with red-and-black ants—and we can't move away from the biting insects because we're fucking pinned down.

The captain gives me a look that says, *Nice job, Wetzel.*

Half an hour later, choppers arrive and surround the area. I ask a gunner if I can fly. If I don't get airborne, that fear I'm always carrying with me is going to take root and ground my courage.

Back at base camp, a rubber tree plantation, I clean my weapons until it's time to go to sleep. I hear a small rustle in the leaves and feel a light striking motion against my boots. I step back and yell for a flashlight.

A guy comes out of a tent and hands me one. In its beam, I see a five-foot pit viper, a venomous snake that can kill you in a heartbeat. And it's getting ready to strike again.

We kill it. When the snake first tried to bite me, its fangs couldn't penetrate my boot leather. I thank the Man Upstairs.

It's nine thirty at night and my balls are dragging. I lie down on my bunk with my boots on and pass out.

A big explosion wakes me up. I stand, about to pull out my .45 when a rocket blows up in a tree and knocks me on my ass. I do a low crawl to the man-made bunker set up between our tents. It's full.

My pants and shirt are covered in shrapnel, but somehow I don't have a single scratch on me. I thank the Big Guy again.

On the eighth of January, I'm up in the air with four other choppers filled with grunts conducting an "eagle flight," a reconnaissance mission designed to simulate the acute vision and fearsome speed of a bird of prey on the hunt.

We land at a place called the French Fort, where we sit around the chopper and eat our late-day C-rations. I take a short walk and check out the prison. Inside, I find the remains of prisoners, the limbs of their skeletons still chained to restraints.

Four Australian EMU choppers land near us. Their ships are shot to shit — damage sustained in an attack about a click away. In a matter of minutes, we're all back up in the air — ten American helicopters and four Australian.

As we make our approach, I see jets conducting air strikes on the opposite riverbank. *Uh-oh. Something's wrong.*

Standard procedure: see your landing zone; fly right by it to conceal your intentions from Charlie; swing around real fast and come in with two sets of gunships. The first gunship preps the landing zone, hitting the area with fire to keep Charlie's head down, while the second moves about one hundred yards ahead of the landing zone.

My chopper is approaching the zone. I locate the gunships about a quarter mile behind us.

Come on, boys. Come on.

We've descended to about treetop level when what seems like two million enemy VC fighters appear on the ground and spring into action. A rocket-propelled grenade makes a direct hit on the left side, where Timmy, our aircraft commander, is piloting the ship.

Now we're coming in *hard*.

In the midst of horrific crossfire, we skid to a landing by the river. One guy jumps out and gets nailed in the head. I'm splattered with what's left of his brains. The front of the chopper's all blown up, so I sling my Thompson submachine gun over my shoulder. I damn near rip the door off as I move through.

My job is to help Timmy. From the waist down, he looks like chopped meat. Bart, my crew chief, helps me unsnap Timmy. We get Timmy about halfway up out of his seat when Bart yells, "Duck," and covers his eyes with his arm.

Fuck. I've got no place to go—

A homemade frag explodes, throwing me against the seat.

"Son of a bitch," I yell. My left arm is damn near blown off— elbow to hand is hanging by some skin.

I grab Timmy with my good hand and give him a little boot in the ass to get him over the radio pedestal. I look to my right, see a VC about to throw another grenade.

I don't know how, but my Thompson is in my right hand. Maybe God put it there. As the VC cocks his arm back, I pull the trigger and zipper him up. He falls back, still holding his grenade—

Boom. A big explosion takes out a few of his buddies.

Shit is blowing up all around me.

Bart drags Timmy out. I step out on the opposite side, fall into the mud, and belly-crawl toward their forward position. Bart and I start putting tourniquets on Timmy's legs to stop the bleeding.

"Look out," Bart says. He has a lot of holes in his head, and he's lost part of his jaw from the shrapnel. "They're coming."

The VC are killing our wounded.

Timmy rolls over. He's all shot up. They're going to think he's dead — Bart, too, given the severity of his wounds. I roll onto my stomach, into the water, and tilt my head just enough to hear and sense movement. When a footfall sounds, I just wait for the bang.

Shoot me in the head and get it over with cuz I'm done anyhow.

I don't know why or how, but the VC shoots me in the foot. It hurts like hell, but the guy doesn't stick around to finish the job. He's run off because our guys are shooting at him. Rifle fire is erupting.

I have a bit more zip left in me, so I grab the Thompson and low crawl around the front of the ship after the shooter and his buddies. The six of them are deep in conversation, trying to figure out how to remove my M60 from the helicopter.

I yell at them. They look at me, startled. I pull the trigger and eliminate them.

When I return, Timmy says, "Tell Jane I love her."

Jane is his wife.

"Shut up, you tell her yourself," I say, even as bullets are zinging everywhere. "We'll get out of this shit."

The VC have disappeared back into the bush. The sun is going down, the tide is coming in, and Timmy is having a hard time keeping his head out of the water. Bart and I slide Timmy across the rice paddy, then Bart takes off, while I use my good arm to support Timmy's head where it's propped on a rice dike.

"Tell Jane..."

Timmy's words trail off, a seeming sign of death.

I don't have time to get emotional. Cry, be sad, whatever. I need to react to the conflict continuing to unfold around me. I need to do my best to survive and help these guys survive.

AMERICAN HEROES

I tuck my wounded arm inside my pants and check my foot where it's been shot. I'm still mobile. Thompson in hand, I do a John Wayne run back to the chopper, because that's where the M60's at.

I'm an air guy, but I can bust a few caps with the best of the grunts. I make it about halfway when I get nailed in the leg. I drop down on one knee. The next thing I know I'm inside the gun well of my helicopter, by my M60.

Come on, you sons of bitches. Come get me. Here I am.

In the chaos, I spot a group of VC. They're armed, but not firing their weapons. By the way they're moving, I can tell they're about to charge the helicopter.

I eliminate them.

I rest until a second wave of VC appear. I eliminate them. More keep coming, their bodies stacking up four or five feet in front of my gun. I keep firing until I'm out of ammo.

I hear screaming and yelling. Not the enemy. Our soldiers.

On the battlefield, people scream and yell for their mamas, their daddies, girlfriends, and their kids. One medic has been shot in the upper back. He's alive, but he can't move. I leave the chopper, armed with my .45. With two gun wounds to the leg and foot, I'm unsteady but still moving. I grab the wounded medic by the shoulder then slide him across the mud, to the rice paddy.

I go back out, bring wounded soldiers to another medic who can patch them up. I grab another guy. Another. I keep doing it until I pass out.

You're a soldier, I tell myself when I wake up. *Help your brothers*. I get my strength back, go out and grab another guy. I keep doing it until I pass out again.

I wake up to a grunt yelling for help. I look around and see he's lying nearby, in elephant grass.

"I'll get you," I say, making my way to him. "I'll get you."

I'm just about to reach him when I startle at an unknown presence to my right. *It's a tiger,* I think, jumping back.

Not a tiger, a VC — and he's about to stab me with a bayonet. Instead of getting me in the guts he sticks me in the right leg and strikes bone.

I pull the trigger. One shot from my .45 puts him out.

War is horrifying. It's not glorifying. Killing is nothing to brag about. It's a part of war, and you deal with it.

I know I'm going to die, but I keep helping my brothers the best I can.

I remove the bayonet, grab the grunt in the elephant grass, and drag him to the medic. I go back out and find another grunt. He has a radio on his back.

"Don't move," I say.

He does — and gets killed. The radio, though, is working. I call in a Mayday. I'm not an artillery guy, but I tell the soldier on the other end of the line where I am, that the VC are on top of us.

Vietnam is hot during the day, but the temperature drops after sundown. I urinate in my pants to stay warm. The sounds of my comrades shouting in fear and pain shatter the darkness.

Twelve hours have passed since the fight started, when, at

sunup, the Army drops troops about a mile from our location. I'm evacuated, I think, on the second helicopter and brought to an evac hospital. Doctors cut me in the arm and both ankles and give me three pints at a crack to keep me from bleeding to death. My heart stops while I'm on the operating table.

I wake up, surprised to be alive. My arm's been amputated, but maybe keeping the partially severed limb close to my body helped save me. Or maybe the Big Guy looked down from upstairs and said, "Keep the redhead alive."

My next duty is in Tokyo. There, my amputated arm gets infected, so I undergo another operation where they remove more bone. I'm grateful, but there's another part of me that feels like a piece of shit for having survived.

Bart is alive. Timmy, I find out, was alive when they placed him on the chopper. He died en route to the hospital.

A lot of death that day. A lot of death. And life goes on.

Three soldiers show up at my bunk. One says, "Are you Gary Wetzel?"

"Yeah."

They all take out their wallets and show me pictures of their wives and kids.

"Because of you," they say, "we get to go back home, back to them."

I want to go back to Vietnam.

"No," the Army says. "You've done your job."

"I ain't done yet."

The Army refuses to send me back.

"Well," I tell them, "if you don't need me, get me the fuck out of here."

I'm given a medical discharge. I spend six months in the hospital. In June, I'm sent back to the States, to the Fitzsimons Army Medical Center in Colorado.

I'm wheeled to the front desk to receive a phone call.

Though Timmy's mom, father, and grandmother can't see me, I stand to greet them.

"Is Janie there?" I ask.

Timmy's wife comes on the line.

I swallow. "Timmy wanted you—"

My throat seizes. I break down. It's like Niagara Falls. Through my tears, I'm able to say, "Timmy told me to tell you he loved you."

Now we're all crying.

Years later, I visit Timmy's gravesite and say the things I've always wanted to say to him. I'm still in touch with his mother, send her flowers on birthdays and Mother's Days.

I turn twenty-one on September 29. For a birthday present, I buy myself a brand-new Corvette. I'm working a new job when a colonel, major, and a first sergeant pay me a visit. They tell me I'm going on an important trip.

A liaison officer fills in the details. I'm going to receive the Medal of Honor. The ceremony will be held at the White House in November.

I don't...this can't be right. It takes the Army about ten days to convince me.

AMERICAN HEROES

* * *

At the White House, I meet other living Medal of Honor recipients, and all I can think is, *What the fuck am I doing here?* I was just a private first class. I'm surrounded by guys I read about in history books.

There's another sergeant receiving the Medal of Honor tonight for his actions in Vietnam. His name is Samuel B. Davis. We get to talking, and then I realize that during the battle, the guy I was talking to on the radio—it was Sam.

I decide to bust his balls a bit. "You know, Sam. Maybe some of these holes I got in my back are from you."

Sam grins. "You wanted the shit dropped on top of you, right?"

"Yeah."

"Well, that's where I put it."

When President Johnson places the ribbon around my neck, I'm extremely flattered and honored to be in the White House with the other four men who are being awarded the same medal.

When I look at the American flag, I see beyond the red, white, and blue. I see what men and women did for me and the rest of us to be here right now. Wearing the Medal of Honor is a privilege, but this medal is for everybody. I'm just a caretaker.

Charles Ritter

Master Sergeant, US Army
Conflict/Era: War on Terrorism (Afghanistan)
Action Date: May 30, 2013
Silver Star

In a place called Lufkin in the heart of East Texas, my mother sleeps with a gun under the pillow. She's worried I'm hanging out with the wrong crowd.

Some of my friends... they're not good people. And truth be told, I'm not a good kid. That said, I'm not a bad person.

Deep down, I know I'm intelligent and resourceful in a particular way that has nothing to do with book smarts. My grades are incredibly bad, but I can always find a way to accomplish anything I'm driven to accomplish.

The year I skipped ninety-plus school days, I went through detention and then into a "special school" for troubled high school kids. And I caused more than my fair share of trouble, racking up seventy thousand dollars in fraudulent credit-card charges from a little scheme I assisted with while working at the Radio Shack store.

I didn't get convicted on any charges, but my parents don't trust me, and they really don't trust my friends, so when they decide to go out of town for the weekend, they lock me out of the house. One of my friends comes over and notices that the kitchen window is cracked open. He manages to get inside, and that's when I decide that instead of staying in the RV parked in our driveway, I'm going to throw a crazy, epic house party.

Everyone is drunk, having a blast. At one point, I go to the freezer and remove a big-ass salmon that becomes the party photo-op. In some pictures, I'm standing in the bathtub with a bunch of girls, everyone holding the salmon. In another, I'm holding a fishing rod and fishing the salmon out of the toilet.

It takes me a day to clean up. I find the salmon in the backyard and return it to the freezer. (Later, my parents will hold an important dinner party and serve that salmon as the main course. The spoiled fish tastes *really* nasty.)

An Army recruiter arrives Sunday morning, along with my parents.

I've always been fascinated by the military. I love reading Tom Clancy books and researching topics like the Cold War and Russian military capabilities and combat tank specifications. I'd spoken to the recruiter before but couldn't do more than toy around with the idea of joining because I still hadn't finished the algebra class I needed to graduate.

"I have some news," he tells me. "Your school has agreed to

waive the algebra class and let you graduate if you join the military."

He shows me my transcripts and my high school diploma. "You can have them when you leave MEPS."

MEPS, I remember, is Military Entrance Processing Station.

"I'll pick you up in front of your house tomorrow morning at oh six hundred."

"Okay," I say. "Sure."

I call a friend after he leaves. "Bring over whatever weed you have," I say. "I have a drug test in the morning that I can't pass."

"Wait. You *want* to fail a drug test? That doesn't make any sense."

I want my transcripts, but I don't want to join the Army. My friend comes over and we smoke weed together. The next day, I go to the MEPS and leave with my graduation certificates.

The following week, the recruiter calls and tells me I failed the drug test. I act like I'm surprised.

My mother refuses to give up on my future. She's a senior nursing instructor at the local community college and manages to get me a full scholarship. I don't excel in any of my classes—including my bowling class. Instead of rolling balls down the lane, I'm out back of the alley smoking weed with the owner. I fail the class and I'm out of college.

I drift deeper into West Texas. Living with my friends, I start to realize I'm going nowhere. I need to do something with my life.

I decide to pursue a military career. My failed drug test is a major obstacle. To get into the Army will require an act of

God — in this case, a letter from a congressman. It's a major pain in the butt, but I manage to get a letter, and I join the Army in 1998. I go in as infantry and choose Hawaii. I'll be posted to Schofield Barracks on the island of Oahu.

In 1991, I was glued to TV coverage of the Gulf War. Every day I'd come home from school and watch the war unfold. I quickly learn the huge difference between watching and doing.

My first PT test in basic training, I score an abysmal 94 out of 300 points. People ask me how it's possible to score so impossibly low. I tell them I took the "you can walk but it's not recommended" part of the briefing to heart and ran a twenty-four-minute two mile.

Fortunately, I have some solid leadership. They take the time to mentor me and help me get squared away. If it weren't for these leaders, I probably would have continued down the path of sucking.

The Army, I realize, enables me to test myself at almost any level. To move beyond my limits to become something more than I am. I end up scoring a 300 on my final PT test.

The day I graduate from basic training, my dad lets me in on a secret.

"Your mother and I had a bet," he says. "I thought there was no way you'd make it through, and she believed you would, thought the Army would be good for you. Looks like I lost."

At Special Forces selection, I show up with a bum knee. I injured it from overstraining.

The O course is at the beginning. I struggle over the next

three weeks covering over one hundred miles while wearing a fifty-five-pound ruck. I finish it, but it's very difficult.

Borderline candidates get "boarded" and that's what happens to me. I came in handicapped with my bum knee, and I wasn't well trained enough on climbing ropes. That's what I tell the commander, but those are not reasons for my failures, only excuses.

"Look, you're a risk," the commander says. "But we're going to select you."

I opt for the 82nd Airborne Division specializing in forcible parachute assault operations. Before I can get to Fort Bragg, North Carolina, 9/11 happens. Two weeks later, the Special Warfare Center for Civil Affairs, Psychological Operations, and Special Forces contacts me.

"You passed Special Forces," the caller says. "We deleted your 82nd orders. You're on orders to start the Q Course."

The Special Forces Qualification Course (SFQC), known informally as the Q Course, is twenty-four months of physically and mentally demanding work. If I make it through, I'll be a Green Beret.

First, I need to go to Airborne School.

I break my ankle on my first jump.

Because of Army regulations, they can't do a "PCS," or Permanent Change of Station. In other words, they're stuck with me. I can't train with a cast, so I'm given charge of quarters (CQ) duty, where I guard the front entrance to the barracks.

Four months later, when the cast comes off, I'm told I have three weeks to complete Airborne School. This time I pass without breaking anything. I make it through the qualification course and graduate in 2003 as a Special Forces weapons sergeant.

I'm excited to receive my duty assignment to Operation Attachment 392. I board a CH-47 helicopter and I'm all kitted out — my body armor and helmet and kit, my old drop leg holster — like I'm fixing for some good combat. I fly to a remote firebase.

The CH-47 lands. The ramp comes down. I see two four-wheelers and four super tanned, shirtless, muscle-bound guys, all wearing Ranger panties and goggles to protect them from the helicopter dust. They've got big beards and long, fabulous hair, and while they're not carrying any weapons, I see, on the perimeter of the helicopter landing zone, four GMV 1.0 Special Operations Gun trucks pulling security and, on the hills, four wheelers with machine gun teams.

"Welcome to Firebase Lane," they say after the bird flies off. "We have two Special Force detachments here, some infantry for perimeter security, and about 120 Afghan Army guys. We've got about a two-hundred-person militia force on our outer perimeter. We're literally in the middle of nowhere, surrounded by people who don't want us here. Oh, and it's really hard for us to get supplied."

From some combination of an outsized ego and "resting dickhead face," I don't make a good first impression. I'm shown to an empty room and given some tools.

Build my bed? I don't know anything about woodworking.

"Can someone help me figure out how to cut this wood and do this stuff?"

The reality is everyone in the company has a Special Forces tab. The piece of cloth I wear on my sleeve is not something that makes me awesome. The only thing the tab gives is a reminder that every single day I need to earn the right to wear it.

I don't learn this right away. Our egos trip us up at times. No one is as great or as bad as we think we are on any given day. Humility, I slowly learn, is king.

Afghanistan looks like the Stone Age, but the Afghan Army is a recent creation. The militia forces are super ragtag but loyal to us because we pay them. We also provide technology. Vehicles, generators, and water pumps. Not much happens—some shooting, some guys lobbing rockets and stuff at us, sure, but nothing super kinetic.

The firebase is brand-new, but it doesn't have any sinks or showers. The nearest water source is a river eight hundred meters away. We go in a group and pull security. I wade into the water wearing fortified sandals because the riverbed is full of sharp rocks and these weird, biting freshwater crabs.

Half our guys jump into the water to wash themselves. We share the river with the Afghans. Their men are in the water, too, shaving their private parts with old-school razor blades.

One day, some guys from the Louisiana National Guard come with us. One guy grabs a bunch of crabs and says, "We're going to eat them."

"First of all," one of the SF guys says, "there are pubes all in that water. Second, don't eat Afghan crab, man. It's just gross."

* * *

I'm sent to Yemen, which is wild. There, people are actively trying to kill us. We have to leave our safe house downtown and consolidate to guard the embassy. We live on the roof pulling sniper overwatch for a terrorist attack. When we eventually pull out, the embassy is bombed the next day.

In 2005, I'm sent to Iraq, where I work with the 3rd Armored Calvary Regiment (ACR) in Tal Afar. My team has brought along a gunfire detection system that uses cameras to show us where a shot came from and the distance it traveled.

I set up the system on top of a castle, the highest point in the city. The first night, we log ten thousand shots. The system is designed to locate sniper rifle from a single shooter, not a city full of thousands shooting at you from all directions. The cameras couldn't keep up, kept spinning in a circle. It's nuts. We decide to turn it off.

We have tremendous firepower. Abrams tanks and Bradley fighting vehicles and unlimited overhead air support. We have a Special Forces company there with an HQ and four teams from the 3rd ACR.

The 3rd ACR takes so many casualties that Colonel H. R. McMaster decides to build a berm around the entire city. Residents who don't plan on fighting are given ten days to leave. Then we start clearing insurgents from the south to the north, block by block.

The enemy, I come to realize, is very good at what they do. They are here to fight, and they're going to face us. They don't care what it costs them.

It's wild. A real eye-opener.

My experience in Tal Afar results in an Army Commendation Medal for Valor. We're put in for much higher awards but the ACR downgrades them.

I'm sent back to Afghanistan for several deployments.

Two thousand eight: Helmand Province. IEDs are such a constant presence that every time we leave the wire, we know someone is going to die. We lose several good people.

In the RG-33—a mine-resistant ambush protected (MRAP) vehicle—we get the powerful protection we need. Its V-shaped hull is designed to deflect blasts with armored glass windows and an automatic fire extinguishing system.

I'm in one of the RG-33's mine-protected seats when we drive over a massive, five-hundred-pound IED. The explosion hurls our vehicle fifty meters down the road. The driver and the Air Force joint tactical attack controller (JTAC) are killed instantly. The captain breaks his back. I'm alive, but I've broken a bunch of ribs. My nose is broken, my right shoulder has some serious issues, and my face is super messed up. A lot of skull fractures and a lot of brain damage, and the whole upper section of my jaw needs to be rebuilt.

While I learn how to walk properly again, surgeons start rebuilding the front of my face. It's a painstaking, four-year process.

I want to deploy again.

People want to know why. Why, after surviving that explosion that killed others, after the long road of rehabilitation followed by years of surgeries and bone grafts and all the pain that came with it—why, after everything I've been through, do I still want to deploy?

I don't have an answer. Even now, I'm still trying to figure that stuff out for myself.

I deploy in 2010 as an E7/SFC, or sergeant first class. I'm a team sergeant on an Operational Department Alpha (ODA). The battalion command sergeant major finds out that I'm not actually cleared for duty (I lied about my medical status), and when we return home, he bans me from deploying until he sees a note from the surgeons and the psychologist that I'm cleared for duty.

In 2013, I return to Afghanistan as a master sergeant. I'm stationed at Camp Morehead, where I'll be working with the 6th Special Operations Kandak.

Kandak is a unit in the Afghan Army that works in cooperation with American forces to bring the hammer and send a message during especially nasty situations. The force ratio stipulation is one Green Beret for every ten Afghans.

We're called to do a company-level mission in a place called Tegab Valley. One Special Forces team is going into the northern valley to push the enemy into the southern part. Forty-eight hours later, a second team will take them out. During the briefing, our small force is told to expect upwards of eight hundred Taliban fighters.

That number seems *way* too high. I don't believe it. Neither do the others.

I'm a part of the southern valley team. My team goes in second. As we land in the mountains under the cover of darkness, the enemy starts shooting at the helicopters. The element of surprise is lost.

A lot of high-level Taliban fighters reside in the area. We secure two battle positions and clear the compounds. It's a long and grueling process that takes all night. We planned on securing three, but the moving through the dark terrain puts us a few hours behind schedule, and now, with the enemy being so close to us, it halts further movement.

There's one main building we can't access. Only a breaching charge will get us inside.

It goes off at three in the morning. We're a kilometer away.

Up until this point, the enemy has been radio silent on comms. Now the radio chatter is off the charts. They're discussing our position and making plans to maneuver on us.

As the sun starts to rise, I'm surprised at how different the terrain looks from the projected imagery. It's densely mazelike and urban. There's no way we can defend it.

We push out multiple small kill teams to push the enemy back. It doesn't work. The Taliban starts hitting us with lots of RPGs. They toss hand grenades over the walls at us. It's a nasty and massive firefight, and it lasts an hour and a half. During that time, we conduct several small kill team patrols to push the enemy back and capture one fighter.

"Hey, Chuck," my team leader calls over the radio, "we're tracking sizeable enemy force. They're southwest of your location and they're moving their way up to you."

A support aircraft starts launching Hellfire missiles as I respond to the team leader. "I see where they are. I think I can get around them to the north. We'll come around and ambush them as they try to get our position. Talk to the commanders."

I'm told the plan makes sense. We move down a narrow alley, only to discover that the enemy has set up multiple machine guns and small arms into a semicircular-shaped ambush.

As we cut around a corner, we encounter a massive amount of fire. The Afghans scatter, running back to their initial battle positions. I hide behind a wall, the only one left.

One of the lead Afghans is down. I can see him through a crack in the wall. Some of his fingers have been shot off and one round went completely through his leg. I can't get around the corner because of the amount of incoming fire.

Going out there alone is suicide. And now there're only two Apache helicopters overhead because the other aircraft left to refuel.

I don't know what's on the other side of this wall. Anytime I peek my head around the corner, I take on intense gunfire. If I can get the Apaches to fire their 30 mm chain gun rounds, it might buy me enough time to grab the wounded Afghan commando.

But the Apaches won't fire. I have no idea why—and I'm getting super pissed. What I don't know is that neither bird has a working weapons system.

An Afghan platoon leader and four other Afghan fighters work their way to my position. We make a decision to get to the wounded commando. There are multiple machine guns out there. It's probably not going to be pretty. Most of us are probably going to die.

It's just what it is. We've got to at least drag this dude around the corner and not let the Taliban get him.

We come around the threshold and fight around the wall of the kill zone, into a wide-open area. I'm in the open, too. There's no cover. It's terrifying.

The enemy is extremely proficient. They fight like we do and have no fear. Two enemy fighters are bounding toward our position. They're within twenty meters, maybe even closer. We shoot them, turn our attention to the machine guns. There's one at each apex of this open area. I toss grenades and go through three mags until all threats are eliminated.

I grab the wounded commando and start dragging him away when a belt-fed PKM machine gun opens fire on our Afghan platoon leader.

I thought I had killed the machine gunner with a grenade. *It must have only rocked him a bit,* I think, transitioning over to where I think the gunner is. I can't see him, but I take a knee and start unloading on his position. I think I'm off target, shooting too far to the left.

The gunner transitions to me and starts firing.

A round hits my leg and knocks me forward. I'm lying there in a somewhat prone Superman position when I take another round to my right upper back. The bullet travels down to my lower spine and, as I will find out later, breaks the brachial nerve complex and the brachial artery. I fall into an Afghan shit ditch as another round slaps me on the butt, the rest of the PKM rounds slamming into the wall behind me.

I'm feeling more stunned than in pain, but my thoughts are

clear, and I'm *pissed*. I need to deactivate that machine gun. It has a one-hundred-round belt. At some point, he's going to have to reload.

The firing stops, followed by silence.

He's reloading. I pop back up to shoot him, but my arm isn't working.

But the Afghans on our team are firing on the gunner's position. I grab our wounded dude with the platoon and say, "Hey, man, let's get round the corner."

Our medic Dan is there. I've lost so much blood that he administers an IV of Hextend, a plasma volume expander. After he gives me a little shot of morphine, I have him reload my rifle for me and then, on my own, move to our casualty collection point. There, an Air Force PJ — a pararescue man and combat medic who often works in the middle of a war zone — tells me I'm bleeding internally.

I'm bundled tight inside a Skedco, which is a litter that looks like a little taco. The team leader (TL) and Dan are going to drag me in the Skedco to the helicopter landing zone (HLZ). It's only going to be the two of them, and they're going to have to fight block by block to get me there because the Taliban is watching us.

The Taliban is speaking over the radio.

"We see them. They're not with the regular Afghans. They've got the wounded guy. We're going to get him. We're going to get him and take everything they have."

Dan and TL stop dragging me and damn near collapse from exhaustion.

"We have to keep moving or they'll kill us," the TL says. "It's just us. If we don't move, we're all dead."

Dan is an incredible soldier, medic, and leader. He and the TL rally and keep fighting street by street to get me to the battle position overlooking the HLZ. With help from an AC-130, they fight the enemy off so the medevac can land. It arrives and takes me away.

Dan and the TL stay behind. They need a way out and speak to the commander.

"The only thing I think we can do—the only thing that's audacious enough to surprise the enemy," the commander says, "is to lay down AC-130 fire on both sides of the road while you move up it."

The road in question is over a mile long and lined with IEDs that are command detonated by a trigger.

"It's the only viable solution," the commander says. "We can't get to you. If you don't do this, you'll be stuck there overnight, and you'll get overrun."

They get ready to run what they refer to as "the Mogadishu Mile." While the AC-130 air support lays down fire on both sides of the road, Dan and the TL toss grenades as they haul ass so fast it takes the enemy completely by surprise.

I crash again on the operating table a couple of times.

My blood pressure is good, but my oxygen levels are dangerously low, and the doctors can't figure out why.

They administer ketamine. It's a "dissociative anesthetic" because the patient is supposed to feel detached from his pain and his environment. But I'm *feeling* the surgery. And I can't move or speak.

A voice says, "I'm watching his vitals. I think he's in pain."

Another voice says, "There's no way he should be feeling any pain. Even if he is, we can't give him any more pain meds because we don't know why he's crashing—or how much morphine they gave him on the battlefield."

They go to work on fixing the artery.

I feel everything.

When I wake up, something isn't right with my back. "It hurts...weirdly," I tell a doctor.

"It's fine. Don't worry."

It's not fine. The artery has popped back open.

They knock me out and fix the artery a second time, but until I get my red blood cell count back up, I won't be stable enough to fly to Germany. It takes five days. Doctors there open me up, remove the bullet, and go to work on repairing my nerves.

I'm sent home with wound VAC therapy. A week and a half later, the doctors close the wound and say, "It's going to take about a year for your arm to recover and work properly."

There's also the issue of my lower back. I'll need a rod and surgery to fuse some of my vertebrae. At some point, I'll also need to get my jaw back into alignment. I assemble a team—a physical therapist, a trainer, and a dietician, all from 3rd Group—and a recovery plan.

"This is going to be my last surgery," I tell them. "My plan is to redeploy."

The group sergeant major calls me. "Chuck, you need to get rid of this pipe dream of redeploying. It's not going to happen."

"It will. If it doesn't, I'm going to get on a civilian flight to Germany. Unless you post a guard at every airport, you can't stop me. I'll take a week's vacation and afterward I'm going to hop on a medevac flight to Afghanistan. I'll see you there."

"You can't come."

"Well, you can go fuck yourself."

Somehow, I don't get fired.

But I do exactly what I said I would do. A month and a half later, I'm back in Afghanistan. The group commander is not happy.

"Well, obviously you made it here," he says. "That's a pretty big testament to our in-group recovery and rehabilitation program. You can go back out to the firebase. I'm allowing it because if we make a big deal about this and you get into trouble, it's not going to look good for anybody.

"That said, you're an idiot. And you can't go out on combat missions anymore."

I end up going out on a combat mission. No one is happy with that, either, but I still don't end up getting fired. Leading an element into fortified machine gun positions overwatching the HLZ, I take a round. They try to medevac me, but I stay in the fight all day and leave at night with the team, after we complete the mission. I receive a Bronze Star for Valor.

After I return from Afghanistan, I do a rotation in Syria and Jordan. When I return home, I take the Combat Physical Fitness test. My hips refuse to cooperate with the run. *Nope,* they say, *we're not doing this one.*

I manage to score a 542. Not exactly what I wanted but it's still above the standards of excellence. I feel like I can perform, but I know I can't perform at the level where I need to be.

It's time to retire.

I'm fine with it. My body is what it is. I suffered brain damage from the explosion, so I do cognitive therapy and some other stuff because my speech gets all weird, but I don't have nightmares. My military service has been a very enjoyable experience. I'm happy with the choices I made, if not with every outcome. I wouldn't change a thing.

There's helmet cam footage from when I was shot. I'm lying there like a sack of shit, in pain, but the response is quick and automated. The medic remains calm, giving orders to the guys around me. They're the ones making the magic happen.

The caliber of people I had the honor of serving with—that, I think, is what drove me to keep coming back again and again. It's been a privilege, not only working with these people, but showing up every day and earning the right to be surrounded by them. To be a part of something more and to do things that are enduring and amazing.

James E. Livingston

Captain, US Marine Corps
Conflict/Era: Vietnam War
Action Date: May 2, 1968
Medal of Honor

I receive my draft notice during my junior year at Auburn University.

In 1961, college students taking a minimum of twelve course hours per quarter can defer. A lot of guys dodge the draft using that technique.

A meeting I have at the student union presents a different route.

A Marine major tells me about the Platoon Leader Class (PLC) Program, which is unique to the Corps. The Junior Platoon Leader Class is designed for candidates between sophomore and junior years, so I'm already a year behind. If I do two PLC boot camps in one summer, I can join the Senior Platoon Leader Class. I decide service to my country is necessary and sign up.

The vigorous mental and physical training doesn't come as much of a surprise. I transferred to Auburn from the Military College of Georgia (now the University of North Georgia). It was a tough little school with an approach comparable to the Citadel. Bad grades? Sit your ass in lock-up. Demerits? Walk them off for three or four hours. I graduated in a high school class of twenty-some students, so my dad sent me to military college to teach me discipline.

It was a good call on my dad's part, giving me that military exposure. My performance during my second PLC boot camp impresses the Marine Corps. I'm offered a regular commission in Quantico, Virginia.

There's one catch. If I accept, I'll forgo the reserve officer uniform allowance. If I don't accept, I'll be given three hundred dollars—a lot of money in those days.

The hell with it. I'm not going to accept the regular commission. I want to take my three hundred bucks and run with it.

I graduate from Auburn University in June of 1962 with a degree in civil engineering. I'm commissioned as a Marine second lieutenant. My plan is to do three years in the service then return to civilian life.

My introduction to Vietnam comes after I cross the Pacific Ocean to Okinawa, Japan. The Marines on board the ship are part of what's called a transplacement battalion. Instead of relieving individual Marines from their deployment, we'll be relieving entire battalions. I'm assigned infantry, which is exactly where I want to be.

From 1963 and well into 1964, I serve mainly off the coast of Vietnam. Sometimes, not a lot, I go into the country. I don't get into any contact to speak of, and when I finish my fifteen months and return to the States, all I keep thinking about is the time I spent with my fellow Marines, how it was an absolute joy. I decide to augment — to become a regular officer.

The Corps ships me down to Parris Island to train recruits and make them Marines. I'm assigned as a series commander, which means I'll have a senior drill instructor and about twenty drill instructors responsible for training about 250 recruits.

I watch the recruits go through the rigorous twelve-week training process, weed out the ones who can't make it. A lot of these recruits were drafted, though in many ways, they're just as sharp and motivated as those candidates who willingly joined. The Corps is constructed as an all-volunteer service, so we're blessed with a lot of these kids saying, "Well, if I've got to go to war, I want to go as a Marine."

Their experiences on Parris Island are so positive that all of them complete paperwork to transition their status from draftee to regular Marine.

Watching these young men come to boot camp, training and pushing them, seeing them graduate and then helping set up their schooling, primarily up at Camp Lejeune, North Carolina, is such a delight that I decide to ditch my three-year plan and reenlist.

I also meet my future wife. We get married in 1966. She's going to live with her parents for two and a half years because I'll be seagoing. It's the start of the Corps' tie-in with the Navy and I've been assigned commander of the Marine detachment aboard the USS Wasp, an aircraft carrier out of the naval yard in South Boston.

Navy capital ships that carry nuclear weapons and have a flag contingent (admiral on board) means Marines are also present to keep the sailors squared away, run the brig, and take care of all the ship's ceremonial needs.

Russians are running up and down our shoreline, and my forty-eight men and I chase after them. I'm really into physical fitness and making sure my Marines are ready to fight, so we also do a lot of training, combat operations, and weapons work at military bases along Cape Cod. My wife and I have only been married a few months. She meets me when she can so we can spend time together along the coast.

After ten months at sea, the ship goes into drydock. I seek out the captain. I'm very frank with him.

"Captain, I'm not going into drydock. I studied engineering in Auburn. If I've got to sit on a ship, I'm resigning my regular commission and leaving the Marine Corps."

I send my resignation letter up through the proper channels. The Navy captain calls Marine Corps headquarters and says, "Send Livingston to Vietnam."

Echo Company, 2nd Battalion, 4th Marines has been in-country since 1965. They've been involved in a lot of fighting, including one of the first big firefights against a major Vietcong force down around Chu Lai, which is south of Da Nang. When I arrive in 1967, the battalion is pretty beat up. They lost a lot of Marines in a major firefight, and the company commander is leaving. I'm replacing him. I'm twenty-seven years old.

I've had my company for about nine hours when I get a mission order.

"We're going to send your company to A Shau Valley," I'm told. "Expect about fifty percent casualties. You have three hours to get ready."

I know how to pull a unit together super quick. I did it a lot at sea. But I don't know these men yet, haven't spent any significant time training with them. I'm going to have to pull my shit together *real* fast.

Fortunately, the operation is called off. Now I can begin the process of rebuilding Echo Company. I need to push these men until I'm sure they've been properly trained.

My leadership style is "do the job." Do the mission we've been sent over here to do, kill the sons of bitches we were sent here to kill. I emphasize daily training: physical fitness, weapons fitness, and weapons cleaning. There's no "winging it." In my company, you don't wing a goddamn thing. That will get these kids killed.

Second, I need to bring these Marines home. That means no slack. It starts at the bottom — the uniform and hair — and goes all the way to the top — the demanding and grueling physical fitness. If a Marine isn't physically fit, he can't fight. And we're going to be fighting. We're going to be playing the game for real.

We're halfway through a river crossing when we take on fire. It's the first time I've taken fire, and all the training we've been through, all the immediate action drills we've practiced, comes into play. Even though we're taking on real bullets, everyone responds correctly. They know exactly what to do.

I'm impressed. The experience also resonates with me to keep training my men, keep pushing.

The Vietcong have no tactical skills, but they do have what I call physical presence skills in terms of their environment. They know every little path, every little nook and cranny, because they grew up here.

The North Vietnamese Army (NVA) is well equipped and has access to plenty of artillery. While they don't have helicopter gunships, naval gunfire, and fixed wing airplanes, they do have a significant number of fighters, and they've been involved in a significant number of battles, the most significant of which is the recent TET Offensive, which occurred only a few months ago.

In late January of 1968, the Vietcong and North Vietnamese launched a series of coordinated attacks in South Vietnam that resulted in brutal and heavy losses for the US and South Vietnamese militaries. Public support for the war has plummeted. Too many American lives, people believe, will be lost to an enemy perceived to be unbeatable.

What no one knows is that the enemy is preparing another bold attack, one it hopes will be the decisive event to fully turn the American public against the war.

On May 1, 1968, Echo Company is defending a bridge on the Cua Viet River when my battalion commander radios me. Dai Do, the nearby village, is overtaken by the enemy.

"We're in a real problem here," he tells me. "Most of the

battalion is engaged, we have lost a significant number of Marines, so we're going to bring you back and have you join the battalion."

Dai Do is *very* close to our Marine combat support base in nearby Dong Ha. The base provides ammo, supplies, and medical support to thousands of our troops fighting in the northern part of North Vietnam. Taking it would be a major victory for the NVA.

We start moving down the Cua Viet River toward my battalion. We have to fight all the way down because the NVA is almost to the point of closing down the river. We take on a lot of fire, especially by snipers, but we don't take any KIAs.

I link up with the remnants of Bravo Company, 1st Battalion, 3rd Marines (Bravo 1/3). Their company commander has been killed. I get the remaining Marines organized.

Golf Company is basically surrounded. We don't know what the size of the contact is, so I call my Marines—particularly the young lieutenants—and give them a mission order.

To get to the village, we're going to have to cross open-air rice paddies spanning five hundred meters. Unlike in the jungle areas, there's no vegetation for cover, so we'll be vulnerable to enemy fire, and I suspect it will be intense. Those of us who make it to the village will then have to deal with another major obstacle. The enemy has set up over a hundred mutually supporting bunkers.

It's going to be a tough fight, but we'll have artillery support from gunships, air, naval—everything that can shoot, we're going to use it. As artillery starts pouring in, I turn to my men. It's 5:00 a.m.

"Guys," I say, "this is the NFL. This is the one we've been waiting on. Fix bayonets."

Every man in the company, including me, is fixed bayonets. The intent is to tell these young Marines that this thing is for real. We've been preparing for this fight for nine months, and we're going to kick ass.

We have no idea that there are nearly ten thousand NVA soldiers in the area. That they've built a massive underground maze with interlocking tunnels. That they have guns strategically positioned at our base in Dong Ha. The enemy is dug in and prepared to fight to the death.

I put two platoons up and one in reserve. The command post, with me and my radio operators, is located between the two attacking platoons.

Dai Do is a typical Vietnamese village, lush with banana and mango trees. The enemy hollows out their interwoven roots and builds low-silhouette bunkers and trench lines so well camouflaged by the ground-level vegetation that they're next to impossible to spot.

As we get closer to Dai Do, we take on intense fire from enemy mortars, rockets, and artillery. Even with the remaining members of Golf Company trying to help us a little bit on the right flank, our attack stalls. The enemy bunkers are proving to be a major obstacle. Nothing less than a direct hit from our artillery even *begins* to do any sort of damage. Anytime one of my Marines gets close enough to fire, he's shot in the head.

It's a battle to the death on both sides.

We're slowing down and momentum is dragging. We've got to take the bunker complex. To do that, we're going to have to penetrate key points and break through.

I move up to the front with my radio operators. I find the weakest point in the bunker complex and order my men to penetrate.

We begin to attack on a very narrow front. The NVA are good fighters, but they're focused on only one direction—across the rice paddy. They can't react to the left and to the right. We're able to penetrate that line of bunkers. Captain Vargas, the commanding officer of Golf Company, and his men see what's going on, jump out from their holes, and start fighting. It gets real nasty, but we're kicking their asses. The remaining NVA flee their positions.

It's nine in the morning. We've destroyed all the bunkers and taken Dai Do.

I started the fight with 180 Marines. Thirty-five are still walking. I have one lieutenant left. The rest of the officers are down. I've taken all kinds of shrapnel and an AK round across the forehead, but I'm not beaten up too bad. The doc wraps me up, pokes a little morphine in me here and there, and I'm back to being functional.

Evacuating the casualties is a major operation. After we get everything consolidated, we redistribute ammo, eat, and drink water.

The regimental commander wants to keep up the pressure on the enemy. He orders the battalion commander to launch an attack on Dinh To, an adjacent village. He sends Hotel Company, 2nd Battalion, 4th Marines around my left flank.

There's about seventy of them. They get about two hundred yards in front of me when I hear them screaming over the radio.

"We're getting surrounded."

"We don't think we're going to make it."

Hotel Company is at risk of being overrun.

"I tell my remaining men, "We don't have any orders, but I'm not going to sit here on my ass while our fellow Marines get wiped out. Does anyone want to come with me?"

They've been through hell and look like they've had all they can handle.

When you're a Marine, you never want to lose a fight. They all volunteer to join me.

We run through intense fire. Smoke and dust. The NVA is throwing everything at us. They're fighting for their lives, and we're fighting for ours. It's close-in killing.

I'm the senior Marine so I sort of take over. We're doing good, knocking the shit out of them. As we drive them back, we seize a series of ridgelines and rice paddy areas. We're doing good, doing good.

As I come across a berm, I see the NVA has one of the old .50-caliber machine guns used to shoot down airplanes. One of the bastards shoots the hell out of me. My radio operator kills the gunner.

I radio the battalion commander. "I'm down. I don't think we're going to survive out here just because of the sheer numbers of the threat against us."

He orders me to collect the wounded and pull back. My men come to pull me out. I tell them to get everyone else first.

"We're getting surrounded," I say when they return for me.

"Y'all are going to get the hell out of here. I'll hold the bastards off as long as I can, until I can't anymore."

"Skipper, we're not leaving you behind."

They drag me away and then move me to the medevac point. The area isn't safe — it's dangerously exposed — but I supervise the evacuation of casualties. I refuse to leave until I know the remaining men on the ground are safe.

As I'm taken away, the regimental commander tells the battalion commander to keep up the pressure. This time, they'll have support from the South Vietnamese ARVN (Army of the Republic of Vietnam) in the form of an armored mechanized battalion with a lot of firepower. The remaining Marines will push the enemy from their strongholds, and once they're out in the open, they'll be destroyed by ARVN tanks.

After the assault begins, the ARVN troops suddenly and without warning disappear — which happens often. ARVN troops are afraid of the NVA, and they include corrupt officers who are in alliance with the enemy.

Wave after wave of fighters surround the remaining 125 Marines. Captain Vargas calls in artillery attacks right on top of them so they can move back. Vargas, wounded himself, keeps going in, again and again, to rescue wounded Marines.

The enemy has made a critical mistake. In their rush to kill the remaining Marines, they've exposed themselves. Fixed wing and helicopters fly in for the kill while the Marines form a defensive line, bracing themselves for a final fight to the death.

The NVA doesn't attack.

As dawn arrives on May 3, the fourth day of fighting, the

Marines discover the NVA has pulled out. They're gone. The battle is over.

If we had lost that battle, lost our Marine combat base in nearby Dong Ha with all its artillery, it would have been, in my estimation, the most significant strategic defeat of the war.

I'm taken to a ship where I receive what will be the first set of operations on my right leg. Fortunately, the round didn't hit bone. If it had, my leg would have been blown off.

I keep thinking of the Marines.

It's difficult to describe to someone who wasn't there, but the way these kids went up against the sheer numbers they were facing while also being concerned about taking care of their buddies and about being a Marine who never wants to lose a fight — these kids went way beyond just doing their job.

When I'm finally able to return to my battalion in Okinawa, I'm told that Captain Jay Vargas and I have been nominated for the Navy Cross for our actions at Dai Do. Shortly thereafter, our awards are upgraded to the Medal of Honor.

When I reflect on being a recipient of the Medal of Honor, my thoughts go to all the young Marines who didn't come home. Men who didn't have a chance to be fathers and grandfathers. I was inspired to become a better Marine because I wanted to represent the values these young men represented. I wear the award for them, and in their spirit, and in their kindness and commitment to the Marine Corps and their country.

Travis Mills

Staff Sergeant, US Army
Conflict/Era: War on Terrorism (Afghanistan)
Date of Action: April 10, 2012
2019 Service Act Honoree for his lifetime of selfless service to members of the combat wounded community.

The colonel wants three infantry guys from Basic to work for his personal security detachment. I'm selected, and in January of 2007, I fly off to Forward Operating Base Salerno, in Khowst, Afghanistan, to start a fifteen-month deployment.

My mom and dad are scared. I'm nineteen and fired up. I went to Airborne School at Fort Benning, learned how to jump out of airplanes, and then started my career with the 82nd Airborne Division. Now I'm a private first class, and I feel ten feet tall and bulletproof.

I work with a very tight-knit group—guys from headquarters, computer analysts, supply people, and mechanics. On TV, a private is always getting yelled at and hazed. Since I'm one of the colonel's guys, nobody bothers me. I'm just one of the guys.

I do patrols with the colonel. There's not a lot of action. What I slowly realize is that when there's danger — like an incident involving a suicide bomber at a nearby hospital, or another incident involving a Vehicle Borne Improvised Explosive Device (VBIED) — a different unit always responds first. Then the engineers clear the area and after that, infantry guys check everything out to make sure it's safe for the colonel.

I never get into a real battle, and not once do I ever fire my gun. Which is crazy, given that I'm in Afghanistan.

I'm friendly with a medic, a married guy from Texas named Josh Buck. In September he leaves Afghanistan so he can head home and see the birth of his daughter. While he's away, I get a MySpace request from a girl named Kelsey.

I don't know her. I click on her profile. She's eighteen and goes to college in Texas. She's also Josh Buck's little sister.

Josh never told me he had a little sister.

Kelsey and I start talking over MySpace. Two months later, we decide we should probably hang out. We book a trip to Cozumel, Mexico, for the last week in December.

After Christmas, I fly to Dallas, Texas, to meet Kelsey's parents. They aren't thrilled about their daughter going on vacation with a stranger. Josh isn't too happy, either, but he knows I'm a good guy and vouches for me.

Spending time with Kelsey, there's no question in my mind that she's "the one."

After Mexico, Kelsey and I head to Michigan, to visit my parents. Then it's time for my second deployment in Afghanistan. I get a layover in Atlanta and call Kelsey. She comes to see me.

"I love you," I tell her.

"I love you, too."

When I return from Afghanistan, Kelsey and I get married.

Our daughter, Chloe, is born in 2011. I'm twenty-four and have just finished a semester at college when I hear my unit is heading back to Afghanistan. The Army says I can sit this one out since I've already completed two combat deployments. I'm ordered to report to Fort Hood, Texas.

I can't stop thinking about my guys.

During my first two deployments, I got a good, up-close look at the people of Afghanistan. They don't care about progress. They're perfectly content using sickles to cut wheat in the fields. I did my best to help them while I was there — building their structures up better, getting rid of the insurgents — but the only thing that mattered to me was making sure the soldier on my left and the soldier on my right would return home safely.

I wasn't there to kill anyone who didn't shoot at me first. I wasn't there to hurt anyone who didn't try to hurt me first. I'm a pretty squared-away person and I'm good at my job, the type of guy you want with you in a firefight. I want to go back again to be there for my soldiers.

I talk it over with Kelsey. While it would suck for me to be away from her and our newborn daughter, getting paid tax-free money plus combat pay and jump pay are strong incentives when you're raising a kid.

My wife understands my decision. I call my sergeant major. He cancels my orders and then I'm off to Afghanistan as a senior squad leader. I'll have forty enlisted guys in my platoon.

We're stationed in a village called Maiwand. It's located fifty miles northeast of Kandahar, Afghanistan's second largest city. Everywhere I look I see mountains and open fields where people live in mud huts and farm by hand, day in and day out. Our forward operating base has no showers. We have to make do using bottled water.

"There's a new standard operating procedure," I'm told by the CO of the former platoon. "You're going to have a guy who's a mine sweeper. He's going to be out in front of you and walking really slow because he's sweeping the ground to see if there are any density changes. And if you get shot at, do *not* dive for cover."

"Okay, so what *do* we do?"

"You take a knee. The enemy plants IEDs in dirt mounds and other places where someone normally dives for cover during an ambush. Also, you don't need to go out all the time and do patrols. The people here don't care about us. You're not going to change their minds."

The first day out on patrol, we're moving through a dry riverbed when we take on fire. Everybody takes a knee as RPGs and mortars and rounds skip all around us. A lot of the new guys are freaked out because this is their first time in battle.

I don't take a knee. My team leaders aren't controlling their rates of fire. I run to each guy, tell them to calm down and control the situation.

First Sergeant Parish reports Sergeant Butler is down. Two soldiers are trying to get Butler to put his arms around them so they can carry him out, but the moment Butler puts pressure on his leg, he falls over.

A squad leader doesn't get into the mix, but I still have the mentality of being the first one through the door, putting myself in danger rather than my men. I put my M4 on safe, hand it to Sergeant Parish and say, "I'll go get him."

I run down the bank. The distance between me and Butler is about the length of a football field. When I reach Butler, lying in the middle of the riverbed, I tell him to climb onto my back. I'm six-three, weigh 250 pounds, and I'm pretty strong from lifting weights. I fireman-carry him to a building about fifty yards behind our firing line.

We get into another big firefight the next day. And the next.

It doesn't ever stop or let up. But we're eating the bad guys alive and gaining ground. I'm always in the mix and getting yelled at for it later because a squad leader isn't supposed to be out there running and gunning.

"It's fine, don't worry," I explain repeatedly. "My guys are trained up. They know what to do if I go down. We're good."

Our rules of engagement are very restrictive. The president of Afghanistan doesn't want us moving around at night and fighting the enemy because it scares his people. Our government agrees, so we can't leave the base at night.

Our raid camera sits thirty feet in the air and can see out to about six kilometers. In the darkness we watch the Taliban put bombs in the ground. We can't shoot them, and we can't leave the base. We just have to try and not walk into those bombs the next day.

On April 10, 2012, we get a call from a village elder about an IED threat in his area. Pretty typical call. The village is next to our strongpoint. This will be a quick mission, two hours, tops.

We go out on patrol — single file, the minesweeping guy in the lead. We come to a short halt to set up overwatch with a pair of 240 Bravo machine guns. We never put them in the same place because the bad guys are always booby-trapping places where they think we'll set up the guns.

The minesweeping guy goes to work, doesn't find anything that alarms him. He marks the area as all safe.

My backpack weighs 120 pounds. It's loaded with ammo, grenades, and a 240. I take my backpack off and place it on the ground, on top of an IED that I'll later learn was made of plastic and copper wire.

I wake up on the ground. On my back. My ears are ringing, and the sun is shining down on me as I lift my left arm. My wrist has been practically severed, but I still have use of my thumb and my index and middle finger, and my first thought is, *I just want to be able to hold my little girl again.*

I roll over. My right arm and leg are gone. My left leg is snapped at the bone, held together by a web of muscle, tendon, and skin.

There's a scene from *Saving Private Ryan* where a soldier gets shot in the stomach and, before he dies, starts crying for his mom and begging not to die. Nothing wrong with going out like that, but I'm not that guy. At the end of the day, I know I'm not in control of the situation. Whatever's going to happen is going to happen.

Don't freak out, I tell myself. *Stay calm.*

My medic and platoon sergeant start putting tourniquets on me. I hear my guys crying out for the medic as well. I recognize their voices—Brandon and Ryan. I ask my medic what happened.

"They're bleeding badly from the face but talking," he says.

"Go save them." I've seen guys who've died from a lot less.

"With all due respect," my medic says, "shut up."

I still have the use of my thumb and at least one working finger, so I grab the trucker mic on my chest and call my lieutenant. "Hey six, this is four. I got guys injured. I need your medic with mine."

The second medic arrives and goes to work on my guys. Then he comes over and helps getting an IV in me. He doesn't know where to put it, so he jams it through my sternum. It's the only pain I feel.

The helicopter arrives within ten minutes. Ryan is crying out in pain from being hit in the testicle. It's bad. I give him a little smile and wink and try to calm him down.

I get the flight mech's attention. "Give my guys water and tell them they're going to be okay."

When we land, I'm taken straight to the hospital. Doctors and nurses are rushing all around me. I keep trying to sit up and—it's really pissing me off—one nurse keeps pushing me down. I need to find my missing leg and arm. I need the doctors to reattach them.

"Where are they?" I ask. "My leg and arm, where did you put them?"

"Honey, they're gone." She looks me up and down and adds, "I don't know how you're still awake right now."

"My baby girl. Am I ever going to see her again?"

It's the last thing I say before the sedation kicks in.

When the doctors wake me up, it's April 14. My birthday.

What's up with me? Why am I hurting so badly?

I'm still heavily sedated. I drift away.

The next time I wake up, Josh, my good friend and now my brother-in-law, is the only one in the room. He tells me he was called and told what had happened. He flew into Kandahar two days ago and has been here since.

"My soldiers," I say. "How are my soldiers?"

He comes to my bedside. "You took the brunt of the explosion for them."

"I can't feel my fingers or toes." I look up at him. "Josh, am I paralyzed?"

"No. You're not paralyzed."

"It's okay. I can take it."

"You're not paralyzed." He speaks slowly and plainly. "But your arms and legs are gone."

"Huh."

I close my eyes and drift away.

The operation, I'm told, took fourteen hours. Nine doctors and seven nurses worked on me. I was given four hundred units of blood, which is the most blood they've ever given to anyone. The blood bank ran out, so a lot of medical staff had donated their own blood for me.

Both legs are gone. My right arm was amputated at mid-bicep. My right leg was amputated above the knee. The left was amputated at the knee. The skin on my left hand had died, so they had to amputate my hand.

Four missing limbs. I'm officially a quadruple amputee.

I had such big plans for my life. I'm a jump master and was planning on going to Fort Bragg to qualify as a "black hat," the term used for Airborne School instructors because of the black baseball hats they wear. I was going to finish my college degree and become an officer. I was going to do twenty years in the military and then retire as a major with a great pension, go out at forty-two years old and choose my next big adventure.

Now that dream is gone. All gone.

How could this have happened? The bad guys weren't supposed to get me. I'm supposed to be the tough guy. The guy who picks up heavy stuff around the house, opens all the jars. The father who picks up his daughter and holds her. That's gone, too.

I shut down. Ignore everyone. The doctors and nurses have questions for me. I look the other way because I'm dealing with my own questions.

Am I a bad person?
Does God hate me?
How can I be a husband and father?
Why didn't I just die?

AMERICAN HEROES

The questions keep coming—and Kelsey keeps calling. I don't answer the phone. I'm deeply embarrassed, can't stop wondering what my life is going to look like going forward.

I don't want to talk to my parents or my wife. Josh finally convinces me to speak to them. I call Kelsey and say, "Hey, what's up? I'm fine. Love you, bye." I get both my parents on the phone, and when my mom wishes me a happy birthday, I hang up.

Three days later I'm flown to Walter Reed. I'm rushed into surgery because the incision in my right leg has ripped open. The doctors are forced to cut higher up on my leg so they can reseal it.

The next day, I see Kelsey for the first time.

I can't look at her.

I've thought a lot about this moment. The way I am now—this shouldn't be her life. She shouldn't have to bear this burden.

"You should leave me," I say.

"Nope."

"You need to go and live a better life."

"That's not how marriage works—how we work."

Chloe is placed on my bed. My six-month-old daughter isn't shocked or repulsed by how her father now looks. All she cares about is that she's with her dad. I can't hold her, but when she crawls onto my chest, I lean my head forward and shower her with as many kisses as I can and think, *I've got to get out of here and get back to my family.*

The pain is so incredibly bad I can't sleep for more than twenty minutes at a time. Every single pain med they give me, the procedures they try—nothing works.

The doctors suggest using ketamine to reset my brain's pain tolerance. They've used it only one time at the hospital. The treatment has only been used thirty times in the world.

I'm given six hundred milligrams of ketamine every hour for five days. I experience crazy hallucinations. I flip out a lot. At one point, I accuse my father of stealing food from a poor family—that's how out of it I am. But the procedure works.

My first week of occupational therapy, I'm all bandaged up and heavily drugged. I've lost 110 pounds. I probably look like a crackhead. I don't know for sure, as I refuse to look at myself in the mirror.

The first few weeks, I feel like a baby. I'm dependent on people feeding me and taking me to use the bathroom and changing my clothes. I can't do anything for myself. This is my life now—and I need to stop dwelling on the unfairness of it all. I need to find a way to move on, to move *forward*.

My wife is recording my sit-ups when the therapist says, "Would you like to take a break?"

I think of my daughter. Chloe and my wife are here with me every day. They're my driving force.

"No," I say. "I'm never going to take a break, never going to quit."

I treat my recovery as a job. Whereas most patients do an hour of physical therapy and an hour of occupational therapy, I do four. That climbs to six and then eight because I'm determined to do something productive each day that will aid my recovery.

I think a lot about the soldiers still serving overseas. A lot of them don't receive care packages. My wife and I send out boxes packed with candy, beef jerky, all sorts of good stuff. It's a small thing, but it keeps me from dwelling on my situation.

When I'm given my new hand, I'm told I can only use it for an hour. I do so well with it because I've worked so hard at getting the muscle flexes right.

"I'm supposed to take this back," my therapist says. "But I'm going to give it to you for a couple of days with the understanding that you can't wear it too much or you'll mess your arm up."

I go to the Fisher House, where my parents and wife are staying, and I'm able to feed myself for the first time — my mom's lasagna.

I learn how to walk at the same time my daughter is learning to walk. The doctors tell me I'll be at Walter Reed for the next three and a half years.

I'm done in nineteen months.

My wife is originally from Maine. We decide to move there. As our house is being built, Kelsey and I move in with her parents, in Dallas, Texas.

Hanging on the wall is an old picture of me taken the day before the IED explosion. I start tearing up, then break down because I'm not that guy anymore. Staff Sergeant Mills, leader of combat soldiers — that guy no longer exists.

So, who am I?

When I was at Walter Reed, I lived in a safety net, surrounded by a group of people who were dedicated to helping me get better. Not only that, the hospital was also full of other guys who had amputations and other injuries due to their service. We were all in the same boat, dealing with the same things. Now I'm in Dallas where nobody looks like me.

And when people see me in the real world, all they see are my wounds. They want to know what's wrong, what happened. I tell them, and while I appreciate their sympathy and compassion, what I really want to say is I'm *not* wounded. I *used* to be wounded. Now, I'm just a man with scars.

Who am I?

I'm not a wounded warrior. I'm a *recalibrated* warrior. My injuries are a part of my life but they're not my whole life.

I start speaking to groups and share my story. A company hires me to speak at their diversity and inclusion conference. The CEO is so impressed with my story and my message that he books me to speak to the whole company. I start getting hired for corporate events.

I keep thinking of other veterans who are like me. A lot of them are suffering and dealing with post-traumatic stress. I'm fortunate to have such a strong support system. A lot of people don't, and they're afraid to ask for help. I've lost some really, really close friends. Their families, their kids and siblings and friends—they would give anything to have them back.

I know I was given a chance to live, move forward and make the most of every day. How can I help them?

The idea comes when I'm on a ski trip with my wife.

The resort, much to my surprise, has rental equipment for "adaptive" skiing, which allows us to ski together. I soon discover there's a whole world of adaptive sports available for people like me. That gets me thinking about how great it would be to bring injured veterans to do activities together with their families.

The Travis Mills Foundation starts as a small camp with little cabins. Our nonprofit provides an all-inclusive, all-expenses-paid vacation to Maine for recalibrated veterans and their families.

My wife and I receive amazing news. We're going to have another child. A boy. We name him Dax, after Daniel and Alexander, the two medics who saved my life.

I'm so lucky to be here. So fortunate.

Who am I?

I'm Travis Mills, a recalibrated veteran and a husband and father to two children. I hate that I got blown up, but at the same time, I don't let that unchangeable fact hold me back.

I live in Maine, where I'm the part owner of an insurance company and a marina. My restaurant, The White Duck Brew Pub, has expanded to accommodate a brewery and an event center.

The Travis Mills Foundation has also expanded. Recalibrated veterans and their families now stay at our new location—cosmetics mogul Elizabeth Arden's former summer estate. I don't take a dime. I'm doing this to give back. It's the right thing.

In 2013, I did an award-winning documentary about my life called *Travis: A Soldier's Story*. I've also completed a second documentary for HBO Max called *Hi, I'm Travis*.

I do about fifty speaking engagements a year. I've written two books: *Bounce Back: 12 Warrior Principles to Reclaim and*

Recalibrate Your Life and my autobiography, *As Tough as They Come*, a *New York Times* bestseller.

My life is pretty fun.

Like everyone, I have bad days. Maybe once or twice a month, right before bed, I go through a "why me" kind of thing where I can't believe I'm going to live the rest of my life without my arms and legs. Maybe it's nighttime that brings it on, but when it happens, I just go downstairs and turn on SportsCenter to stop the dark thoughts from creeping in.

Two things I've learned keep me going and keep me sane.

First, I don't dwell on the past. I can't change yesterday or what happened eleven years ago. Instead of dwelling, I reminisce. For my first twenty-five years of life, I had arms and legs. Those were incredible years. As I was lying in my hospital bed on my twenty-fifth birthday, if you told me I was going to be successful the way I am right now, how successful and lucky and fortunate I would feel—that I had more incredible years ahead of me—I would have told you there's no way.

Second, I can't always control my situation, but I can always control my attitude. No matter how pissed or angry I get, no matter how much I want to yell, it's not going to fix anything.

And then there's the future. Thinking about it can drive me crazy. How am I going to teach my son how to play baseball? How am I going to dance with my daughter?

I have no idea. But I do know I'll find a way.

Ernest E. West

Private First Class, US Army
Conflict/Era: Korean War
Action Date: October 12, 1952
Medal of Honor

Interview with Amy West Hogsett, daughter

An enormous empty field stands in front of our house in Wurtland, Kentucky. Sometimes a helicopter lands there to pick my father up.

I know my father as the high school football and basketball coach. I have no idea he's a war hero.

Some weekends we drive to the Army bases in Fort Knox or Frankfort, Kentucky, where my dad does military work, conducting entrance and exit interviews with troops.

I don't think much about it. It's just a part of my life, just like going to PTA meetings and working at the high school games and the fall and spring carnivals.

My parents are great people, soulmates, who have the best relationship on the planet. They met after my father left active service. My mom was working at the post office then, sorting mail. She was twenty-seven when they got married. They lived a lot of life in the ten years before I, their only child, arrived. Now she's a schoolteacher, and they're very involved in the community. They refuse to allow a single person—a child, especially—to want for anything.

As I get a little older, I see my dad get more political. He's involved with the State of Kentucky—and really involved with the Greenup County War Memorial in Wurtland. He sketched the whole thing out then raised money to bring over a WWII Navy Landing Craft, a decommissioned UH-1 Huey helicopter, an M60 battle tank, and an F-86 fighter jet. Etched into a huge piece of granite are the names of every person from Greenup County who lost their lives in war.

Greenup is the only county in the US that has two Medal of Honor recipients. My dad, I discover, is one. The other is John Collier, who died in the Korean War. He threw himself on a grenade to save the lives of the soldiers in his unit. My dad talks about him at every event. They once worked side-by-side on the Chesapeake and Ohio Railroad.

What I learn about my dad's service in Korea comes from hearing him speak, often at the local middle school and high school, and give interviews. The details never come up at home except when my mother tells me things. He's very humble, doesn't consider himself a war hero. When anyone compliments him, he says, "It's not about me. I came home. We were all there, even people I didn't know, and a lot of us didn't come home. That's who it's about."

Ernest Edison West, the man who would later become my father, grew up in the Depression. His parents had a lot of kids and didn't take care of them. My father and his sister were neglected, to the point where every day they would have to go out and search for food. They had a lot of success climbing trees and finding bird eggs to eat. Finally, the local church called the Methodist Orphan's Home in Versailles.

"I was raised with 125 brothers," he said. "We all stuck together."

In 1950, when he got drafted, he felt it was his obligation to go and serve his country. "I didn't even know where in heck Korea was. Didn't even know we were in a war until we started reading in the paper about it."

On October 12, 1952, Private First Class West was up on Heartbreak Ridge when he volunteered to join a combat patrol near Sataeri, Korea. They had spotted a bunker and watched it for three days. My father was the point man for the three-squad patrol.

"We started up the hill and all hell broke loose."

The enemy started lobbing grenades down the hill, Private West watching as they rolled between his legs and exploded, taking out the men behind him. One exploded in front of him.

A few minutes later, after he regained consciousness, he saw the enemy coming up over the hill. "It's frightening because it looked like a million people coming right at you. It was...Lord behold. You just sat there and fired until you can't shoot no more."

A lot of his fellow soldiers were wounded, including their leader, Lieutenant George M. Gividen. Private West told everyone to withdraw, then went back alone to retrieve the lieutenant.

"I got into a firefight with George over my shoulder. I had two carbines, one in each hand. Firing them. And they started firing at me, so I dropped George on the ground and laid on top of him. And then I got up and killed the guys coming after us."

After he delivered George to the others, Private West noticed three men were missing. He told everybody to hold their position.

"I'll be back. We will all be going back home."

He succeeded in saving all three of his fallen brothers.

"I had shrapnel in my eye. I was burnt. I was shot. And to be honest with you I didn't know none of it had happened. I didn't think I could get hurt, I guess."

The shrapnel wound cost him his eye. He spent ten months in the hospital and then returned home to Kentucky. One day, when he went to collect his mail at the post office, he learned he had a telegram. It informed him that he was receiving the Congressional Medal of Honor.

He turned it down.

"I thought if one was going to get the medal, everybody ought to have one. We all went, we all served—if you give one, you ought to give one to everybody. That's the way I feel about it—but they don't."

I have a lot of stories about my dad, but one of my favorites is about when he was the mayor of Wurtland. The town needed a new sewer and water system, and he worked very diligently with the State of Kentucky to help get it.

Many local people are without city water, and it costs money to connect it from the road to the house. It's the property owner's responsibility to dig the trench and lay the pipe. My dad would work all day at CSX Railroad (formerly Chesapeake and Ohio Railroad), change and eat, and then go out to help people dig their water lines.

He did this day after day. For years.

The Wurtland Middle School gymnasium is named after my dad. We had his funeral there. They honor him on Veterans Day.

My whole life, Dad talked about the word "brotherhood." How, when he was living in the Methodist Orphan's Home, he had 125 adopted brothers who looked after him and protected him. He did the same for them and for his brothers in the Army. It was, he always told me, the way he was raised.

"I think today, in this day and time, if we would think of brothers and sisters, that we'd have a better society to live in. We need to be good to people. Treat people like you'd like to be treated. We're all the same. We might have a little different accent, but we're all the same. We're Americans."

Acknowledgments

For their invaluable assistance and expertise, the authors recognize George Bennett; Caroline Hirsch, Stand Up for Heroes, Bob Woodruff Foundation; and Neil Thorne, Veterans Advocate.

At the Army Women's Foundation, thanks to Anne Macdonald, Brigadier General US Army (Ret.), President; and Beth Spitzley, Director of Administration.

Very special appreciation goes to the Congressional Medal of Honor Society.

Bibliography

The authors relied on many sources in preparing this book. Each of the chapters in this book is based largely on an interview with the protagonist of the story or, if he or she is no longer alive, someone close to him. The authors consulted other accounts of the action for which a person was honored in the books and articles listed here.

Hershel "Woody" Williams (Brent Casey)

Michael E. Ruane, "A farm boy became a fearsome warrior at Iwo Jima. And he did it with a flamethrower," *Washington Post,* February 18, 2020.

Emily Langer, "Hershel Williams, Last Medal of Honor Recipient from World War II, Dies at 98," *Washington Post,* June 29, 2022.

Brent Casey, "Brent Casey: A Veteran's Story," US Department of Veteran Affairs.

Hershel Williams, "Last Surviving WWII Medal of Honor Recipient Hershel 'Woody' Williams' heroic actions on Iwo Jima," American Veterans Center.

Hershel Williams, "Destroying Pillboxes with a Flamethrower on Iwo Jima," American Veterans Center.

BIBLIOGRAPHY

Blake Stilwell, "This Is the Marine Corps Flamethrower Tank That Won at Iwo Jima," Military.com, February 17, 2021.

David Vergun and Katie Lange, "Iwo Jima Medal of Honor Recipient Recounts Battle Experiences," DOD News, November 8, 2021.

"In the Footsteps of Hershel 'Woody' Williams," Medal of Honor Valor Trail.

Joe Archino, "Bravery on the Black Sands: The Heroism of Hershel 'Woody' Williams at the Battle of Iwo Jima," thisiswhywestand.net.

Patrick Henry Brady

Patrick Brady, Major, US Army, "Near Chu Lai, South Vietnam, 1968," video produced by Medal of Honor Book, youtube.com/@MedalOfHonorBook.

Patrick Henry Brady, "Medal of Honor: Landing in a Minefield to Save 51 Men," American Veterans Center.

No Byline, "Charles L. Kelly," Hall of Valor: The Military Medals Database, valor.militarytimes.com.

Paris D. Davis

Drew F. Lawrence, "Fire Watch: Col. Paris D. Davis and His Decades-Long Wait for the Medal of Honor," Military.com, March 10, 2023.

Haley Britzky, Oren Liebermann, and Betsy Klein, "Special Forces soldier who 'never' quit receives Medal of Honor nearly 60 years after grueling firefight," CNN, March 3, 2023.

Dave Philipps, "A Black Soldier's Heroism, Overlooked in 1965, May Finally Be Lauded in 2021," *New York Times,* February 15, 2021, updated November 3, 2021.

Matt White, "'I Just Disobeyed the Order: The Incredible Story of Capt. Paris Davis' Medal of Honor," Coffeordie.com, February 14, 2023.

David Roza, "How Green Beret Paris Davis' teammates fought the Pentagon for his Medal of Honor," TaskandPurpose.com, February 21, 2023.

Dustin Jones and Devin Speak, "Decades after risking his life to save his men, a Green Beret gets the Medal of Honor," NPR.org, updated March 3, 2023.

BIBLIOGRAPHY

President Joseph Biden, "Remarks by President Biden at Presentation of the Medal of Honor to Army Colonel Paris Davis," WhiteHouse.gov, March 3, 2023.

Catherine Herridge, "A Black Vietnam War veteran was nominated for the Medal of Honor. He's still waiting 56 years later," CBSNews.com, updated on April 2, 2021.

Catherine Herridge, "Black Vietnam veteran's nearly 60-year wait for Medal of Honor is over," CBSNews.com, updated on February 14, 2023.

Matt Pusatory, "Arlington soldier to receive Medal of Honor for bravery during Vietnam War," wusa9.com, March 2, 2023, updated March 3, 2023.

Thomas William Bennett (George Bennett)

Cynthia Mullens, "West Virginia Veterans Memorial," West Virginia Archives and History, archive.wvculture.org, December 2017.

Jim Bissett, *The Dominion Post*, Morgantown, W. Va., "Biography of Morgantown Medal of Honor recipient Tom Bennett republished," Yahoo.com, May 28, 2023.

History.net Staff, "A Conscientious Objector's Medal of Honor," History.net, June 12, 2006.

Cindy Pritchett

"Living Legend—CSM Cindy Pritchett, USA Ret.," US Army Heritage & Education Center, ArmyHeritage.org.

"Transcript: Interview with Cynthia Pritchett," "Unsung Heroes: The Story of America's Female Patriots," US Army Educational Package, Unsung HeroesEducation.com, November 26, 2012.

"Command Sergeant Major Cindy Pritchett—Afghanistan," US Army Heritage & Education Center, ArmyHeritage.org.

Ralph Puckett, Jr.

Ralph Puckett, Jr., "Living History of Medal of Honor Recipient Ralph Puckett, Jr.," Congressional Medal of Honor Society.

BIBLIOGRAPHY

Ralph Puckett, Jr., "Medal of Honor: Colonel Ralph Puckett Jr.," US Army, army.mil.
Richard Pearson, "Army Col. Louis Mendez Jr.," *Washington Post,* September 23, 2001.
Devon L. Suits, "Medal of Honor: Korean conflict hero led Rangers in battle for Hill 205," Army News Service, army.mil, May 20, 2021.
"Medal of Honor: Colonel Ralph Puckett Jr.," US Army, army.mil.
Dr. Michael E. Krivdo, USASOC History Office, "The Battle for Hill 205: US Army Rangers and the Beginning of the Korean War's Third Phase," US Army, army.mil, December 4, 2020.
Glenn Thrush, "Biden awards the Medal of Honor to a Korean War hero," *New York Times,* May 21, 2021.
"Ralph Puckett," Hall of Valor: The Military Medals Database, valor.militarytimes.com.

Duane Edgar Dewey (Arline Broome)

Stephen Brooks, "Vietnam veteran receives honorary diploma 46 years after high school's graduation," MLive.com, July 13, 2014.
Duane Dewey, Colonel, US Marine Corps, "Panmunjom, Korea, April, 1952," video produced by Medal of Honor Book, youtube.com/@MedalOfHonorBook.
Harrison Smith, "Duane E. Dewey, Medal of Honor recipient with 'a body of steel,' dies at 89," *Washington Post,* October 22, 2021.

Alwyn Cashe (Tamara Cashe)

Dan Lamothe, "After extraordinary sacrifice — and years of delay — Alwyn Cashe gets his Medal of Honor," *Washington Post,* December 15, 2021.
"Army Sgt. 1st Class Alwyn C. Cashe — Honor the Fallen," thefallen.militarytimes.com.
Sgt. 1st Class Justin A. Naylor, "A Dogface Soldier through and through: Memories of SFC Alwyn Cashe," US Army, army.mil, December 13, 2021.
David Zucchino, "Medal of Honor campaign continues for black sergeant who saved troops," *Los Angeles Times,* December 7, 2014.

BIBLIOGRAPHY

Michael J. Tarpey, "Selfless and courageous: Surgeon remembers soldier chosen for Medal of Honor," MilitaryTimes.com, December 15, 2021.

Harvey Curtiss "Barney" Barnum, Jr.

"Living History of Medal of Honor Recipient Harvey Curtiss 'Barney' Barnum, Jr.," The Congressional Medal of Honor Society.
Andrew McClain, "Harvey Barnum and Operation Harvest Moon: December 18, 1965," VA News, news.va.gov, December 18, 2022.

Matthew O. Williams

"Living Medal of Honor Recipient Matthew Williams," The Congressional Medal of Honor Society.
Devon L. Suits, "Shok Valley weapons sergeant to receive Medal of Honor," US Army, army.mil, October 11, 2019.
Adam Linehan, "Unsung Heroes: 12 Green Berets Who Took on Hundreds in This Epic Afghan Battle," TaskandPurpose.com, July 28, 2015.

Donna Barbisch

"Remembering Donna Barbisch," Circulating Now, circulatingnow.nlm.nih.gov, April 26, 2018.
"Women Veterans' Stories of Service: Donna Barbisch," Veterans Health Administration.

David G. Bellavia

Jeff Edwards, "Fighting Through the Streets of Fallujah, David Bellavia Survived Brutal Hand to Hand Combat," War History Online, WarHistoryOnline.com, April 18, 2016.
"How one soldier wiped an entire enemy squad in Fallujah," Task & Purpose.
"Medal of Honor: Intense Close-Quarters Firefight in Iraq | David Bellavia," American Veterans Center.

BIBLIOGRAPHY

Michael Ware, "Into the Hot Zone," *Time,* November 22, 2004.
President Donald J. Trump, "Remarks on Presenting the Medal of Honor to Staff Sergeant David G. Bellavia," The Presidency Project, presidency.ucsb.edu., June 25, 2019.

Chuck Pfeifer

Chuck Pfeifer with Ivan Solotaroff, "The Ballad of a Green Beret," *Esquire,* September 1994.
Spencer Morgan, "The Wild Purple Heart of Chuck Pfeifer," *Observer,* January 6, 2009.
No Byline, "Attack on FOB 4: The Worst Day in US Army Special Forces History," Coffee or Die, CoffeeOrDie.com, October 24, 2021. The article was originally published on September 28 on Sandboxx News.

Paul Zurkowski

Tech. Sgt. David Speicher, "Airmen awarded Distinguished Flying Cross," Air Force, af.mil, December 12, 2013.
"The Warthogs Stunning Rescue Mission in Taliban Territory," Smithsonian Channel Aviation Nation.

Mark Mitchell

"Afghanistan 'horse soldier' author visits Edgar school," *Star News.*
Dodge Billingsley, "Insurrection at Qala-i-Jangi," *The Harriman Review,* Columbia University Libraries.
Dan De Luce, Brenda Breslauer, and Yasmine Slam, "A CIA veteran who survived a hand-to-hand battle with Al Qaeda is now helping Afghans escape the Taliban," NBC News, NBCNews.com, November 11, 2022.
Video, "Qala-i-Jangi Prison Uprising, US-UK Special Forces, November 2001," Liveth For Evermore.
Lauren Katzenberg, "When It Comes to Bowe Bergdahl, 'We All Really Failed,'" *New York Times,* March 15, 2019.

BIBLIOGRAPHY

PBS NewsHour, "Why did Bowe Bergdahl leave his post? Army transcript sheds light," PBS.org, March 16, 2016.

Sandboxx, The Military Platform, "First Casualty: A Gripping Account of the CIA in Afghanistan after 9/11," Sandboxx.us, undated article.

Conrad Begaye

Kent Harris, "Vicenza NCO receives Silver Star for saving lives in Afghan ambush," *Stars and Stripes,* July 1, 2009.

Capt. Joseph Sanfilippo, "NCO awarded Silver Star for courage under fire in Afghanistan," US Army, army.mil, October 23, 2012.

McKenna "Frank" Miller

Pilot Online, "Green Beret from Virginia Beach receives Silver Star," *The Virginian Pilot,* January 9, 2012, updated August 7, 2019.

Tim Sheehy

Thom Bridge, "Sheehy Bronze Star," *Independent Record,* helenair.com, August 27, 2015, updated October 4, 2015.

Jay Zeamer, Jr. (Barbara Zeamer)

Nicole Morell, "Remembering a Hero—Veteran Jay Zeamer Jr. '40, SM '45," MIT Alumni, alum.mit.edu, November 10, 2014.

No Byline, "Jay Zeamer Jr.," Stories of Sacrifice, The Congressional Medal of Honor Society.

Los Angeles Times Archives, "Jay Zeamer Jr. 88, pilot won the Medal of Honor in World War II," *Los Angeles Times,* March 26, 2007.

Earl D. Plumlee

Tim Sileo, "Having Their Backs," The Stream, stream.org, January 27, 2015.

BIBLIOGRAPHY

No Byline, "Michael Harrold Ollis," Hall of Valor: The Military Medals Database, valor.militarytimes.com.

Dan Lamothe, "Soldier at center of Medal of Honor controversy recalls the day he faced suicide bombers," *Washington Post*, June 27, 2016.

No Byline, "Master Sergeant Earl D. Plumlee, Medal of Honor, Operation Enduring Freedom," US Army, army.mil.

Sgt. Michael Sword, "Plumlee earns Silver Star," US Army, army.mil.

Mustafa Andalib, "Taliban attack Polish base in Afghanistan, seven killed," Reuters, August 28, 2013.

Alan Mack

Interview, "160th Special Operations Aviation Regiment at War, Alan Mack, Ep. 199," The Team House.

Interview, "Alan Mack Helicopter pilot shot down twice during Operation Anaconda," Shots from the Winchester.

Interview, "Legendary Army Pilot Reveals His Experiences in Afghanistan," OutKick.

Razor 03: A Night Stalker's War by Alan C. Mack, pages 126–130, Pen & Sword Military, publication date September 15, 2022.

Gary Wetzel

"Living History of Medal of Honor Recipient Gary Wetzel," The Congressional Medal of Honor Society.

"'Eagle Flights' prey on fleeing Viet Cong," *Stars and Stripes*, May 1, 1967.

Interview, "Gary George Wetzel Collection," Library of Congress.

James Livingston

Katie Lange, "Medal of Honor Monday: Marine Corps Maj. Gen. James Livingston," *DOD News*, May 2, 2022.

"Magnificent Bastards of Dai Do," Against the Odds, Season 2, Episode 3, American Heroes Channel, air date February 29, 2016.

BIBLIOGRAPHY

"James Livingston, Medal of Honor, Vietnam War," video produced by Medal of Honor Book, youtube.com/@MedalOfHonorBook.

Travis Mills

Travis Mills Foundation, travismillsfoundation.org.

Fotolanthropy (producer), Jonathon Link (director), Jonathon Link (writer), *Travis: A Soldier's Story,* documentary.

Jon Michael Simpson (director), Jon Michael Simpson and C. Bailey Werner (writers), *Hi, I'm Travis Mills,* TV series, directed by Jon Michael Simpson, written by Jon Michael Simpson and C. Bailey Werner. Aired? November 11, 2022.

ABC News, "Michigan 'Hero's Welcome' for Staff Sgt. Travis Mills, Who Lost Four Limbs," ABC News, October 4, 2012.

US Veterans Magazine, "Travis Mills Profile in True Courage," *US Veterans Magazine,* October 2020.

Ernest E. West (Amy West Hogsett)

Tim Preston, "A stamp of approval for war veteran Ernie West," *The Daily Independent,* August 4, 2014.

"Remembering Ernest E. West," National Medal of Honor Museum, May 5, 2021.

"Ernest West, Medal of Honor, Korean War," video produced by Medal of Honor Book, youtube.com/@MedalOfHonorBook.

"Medal of Honor Recipient Ernest E. West Passes Away at 89 Earned Nation's Highest Award for Valor during the Korean War," Congressional Medal of Honor Society, May 3, 2021.

About the Authors

James Patterson is one of the best-known and biggest-selling writers of all time. Among his creations are some of the world's most popular series, including Alex Cross, the Women's Murder Club, Michael Bennett and the Private novels. He has written many other number one bestsellers including collaborations with President Bill Clinton, Dolly Parton and Michael Crichton, stand-alone thrillers and non-fiction. James has donated millions in grants to independent bookshops and has been the most borrowed adult author in UK libraries for the past fourteen years in a row. He lives in Florida with his family.

Matt Eversmann retired from the Army after twenty years of service. His first book with James Patterson was *Walk in My Combat Boots*.

Tim Malloy is a veteran of print and television journalism. He has won fourteen Emmys as an investigative reporter, documentary maker, and war correspondent.

Chris Mooney is the international bestselling author of fourteen thrillers. The Mystery Writers of America nominated *Remembering Sarah* for an Edgar Award. He teaches creative writing at Harvard.